I0488172

21st CENTURY
SCIENCE & TECHNOLOGY

Special Anthology // 150 Years of Vernadsky: The Noösphere (Volume 2)

150 YEARS OF VERNADSKY (VOLUME 2):
THE NOÖSPHERE

Contents

EDITORIAL STAFF

Editor-in-Chief
Jason Ross

Managing Editor
Marsha Freeman

Associate Editor
Christine Craig

Staff Writers
Megan Beets
Benjamin Deniston
Liona Fan-Chiang
Creighton Jones
Michael Kirsch
Natalie Lovegren
Meghan Rouillard

21st Century Science & Technology (ISSN 0895-6820) is an occasional publication by 21st Century Science Associates, 60 Sycolin Road, Suite 203, Leesburg, VA 20175. Tel. (703) 777-6943.

Address all correspondence to: 21st Century, P.O. Box 16285, Washington, D.C. 20041.

Letters to the editor: letters@21stcenturysciencetech.com

21st Century is dedicated to the promotion of unending scientific progress, all directed to serve the proper common aims of mankind.

Opinions expressed in articles are not necessarily those of 21st Century Science Associates.

We are not responsible for unsolicited manuscripts.

www.21stcenturysciencetech.com

Vladimir Vernadsky
Scientist for Mankind's Future

March 12, 2014 marked what would have been the 151st birthday of the great Russian-Ukrainian scientist Vladimir Vernadsky. This past year was dedicated in Russia as a celebration of his life and work, an occasion which was also honored by *21st Century Science and Technology*, including our participation at an event at the Ukrainian embassy, as well as the release of several new and original English translations of Vernadsky's work, including a new and original English translation of a fascinating speech delivered by Vernadsky in 1931, entitled "On the Conditions of the Appearance of Life on Earth." Other original translations previously published by *21st Century Science and Technology* over the years have been compiled for this Vernadsky anniversary edition, including "The Evolution of Species and Living Matter," "Human Autotrophy," and excerpts from Vernadsky's "Scientific Thought as a Planetary Phenomenon."

Throughout the work of Vernadsky, we find a powerful argument for why processes on Earth, and in the Universe, are organized according to a top-down principle of life, and, even higher, human cognition. This is a concept found throughout the writings and speeches of economist Lyndon LaRouche, who has often referenced the work of Vernadsky.

The work of Lyndon LaRouche has focused more explicitly on the fundamental distinction between animal and human life: although the physical appearance of man and ape might have a striking resemblance, the actions able to be enacted by those two species are entirely different, and they represent a specific kind of difference. The difference is not a matter of choice, or of taste: animals are fundamentally incapable of creating the kinds of changes to themselves as well as their environment that human beings are capable of, and there are quantitative effects which illustrate this, central to Mr. LaRouche's science of physical economy. For example, animal population growth, which may appear at first to be more rapid when examining a relatively short interval of time (as in the well-known case of rabbit population growth), will ultimately level out, as the animal populations are unable to make any fundamental changes to their environment to increase its potential habitability.

Vernadsky's life's work ended up culminating in a similar investigation, of the unique distinction of man from animal, something Vernadsky approached from the standpoint of a biogeochemist. Having immersed himself in the study of how life transformed the surface of the Earth, he marveled at the power of human creativity to do the same, but on a quantitatively and qualitatively higher level. In his paper "Some Words on the Noosphere," Vernadsky asked:

How can thought change material processes? Here a new riddle has arisen before us. Thought is not a form of energy. How then can it change material processes? That question has not as yet been solved… as for the coming of the noosphere, we see around us at every step the empirical results of that "incomprehensible" process. That mineralogical rarity, native iron, is now being produced by the billions of tons. Native aluminum, which never before existed on our planet, is now produced in any quantity. The same is true with regard to the countless numbers of artificial chemical combinations (biogenic "cultural" minerals) newly created on our planet.

Vernadsky also expressed his confidence in the noösphere, despite the World War which he found himself in the middle of, saying:

At present we cannot afford not to realize that, in the great historical tragedy through which we live, we have elementally chosen the right path leading into the noosphere. I say elementally, as the whole history of mankind is proceeding in this direction.

Today we find ourselves in a similar situation, faced with the threat of a great new war. The only solution to the crisis faced by mankind lies in the further development of the noösphere as Vernadsky saw it.

Vernadsky was a patriot of both Ukraine and Russia, and would surely look upon the current situation in this region with both sorrow and disdain, but also, hope, confident that the only future for both nations, and the rest of the world, lies in the domain of a commitment to unending scientific and technological progress, a requirement which is incompatible with the thinking of many citizens and world leaders today, but a commitment which *21st Century Science and Technology* is actively fighting for.

To this end, we are publishing this two-volume set of selections of Vernadsky's work, and critical essays on its implications. In Volume I, we feature Vernadsky's writings, and accompanying articles, which focus more on his concept of the biosphere, and living matter and its distinction from non-living matter. In Volume II, we feature those writings and articles which deal more explicitly with the power of human cognition as a geological force: the noösphere.

May the thoughts of Vernadsky find their rightful place in science, and politics, today.

Meghan Rouillard, Series Editor

The Transition From the Biosphere To the Noösphere

by Vladimir Vernadsky

Excerpts from
*Scientific Thought as a
Planetary Phenomenon*
1938

Translated by William Jones

*Vladimir Ivanovich Vernadsky
1863-1945*

Introduction

by William Jones

The name of Vladimir Ivanovich Vernadsky may be familiar to many people involved in the area of science, particularly in the geological and so-called "earth" sciences, but most of these scientists, without a good working knowledge of Russian, will only have known his work, at best, through the publication of his 1926 monograph, "The Biosphere," which brought him some immediate international attention since it soon appeared (in 1929) in a French edition. This has since been translated into many languages, although first appearing in English only in 1986. Since the 1980s, the work of Vernadsky has been widely circulated and popularized by the movement led by U.S. economist and statesman Lyndon LaRouche, whose work on economics has, over the last few decades, been most significantly influenced by Vernadsky's concept of the noösphere. In their view of man and man's possibilities for development they are kindred souls.

In Russia, Vernadsky's name is as familiar as that of Pasteur or Curie or Einstein. President Vladimir Putin has decreed that the 150th anniversary of Vernasky's birth next year will be the occasion for celebration throughout the country. While much of Vernadsky's early work first appeared in French scientific journals, most of his major works, including his last, unfinished, magnum opus, "The Chemical Structure of the Biosphere and Its Surroundings," exists only in Russian. In fact, since Vernadsky, working for the first part of his life under the Tsarist regime and the last part under the Soviet regime, was in both cases considered something of a "dissident," many of his most path-breaking and creative works were not published until well after his death.

Vernadsky's life covers a long and dramatic span of Russian history. Born in 1863 in the midst of the great reforms initiated by Alexander II and living until the very eve of the end of World War II, dying in January, 1945, Vernadsky was an active participant in some of the greatest upheavals of that era. Born in St. Petersburg, he spent much of his early life and young manhood in Ukraine, the family having its roots in that region.

Studying during one of the most fertile periods in Russian science under the great chemist Dmitry Men-deleyev, and the renowned soil scientist V.I. Dokuchaev, Vernadsky was first drawn to the study of crystallography and mineralogy. Vernadsky went on expeditions with Dokuchaev to study the fertile "black earth" of Ukraine, where his attention was first attracted to the elements of living organisms that contributed to that soil's tremendous productivity. Indeed, it would be later, during his temporary exile in Ukraine after the Bolshevik Revolution that Vernadsky would first develop his own unique concept of the role of the "biosphere."

But Vernadsky, like Leonardo da Vinci, one of his great heroes in the realm of science, was also something of a universal genius. His interests spread over the entire gamut of scientific thought. And like Leonardo, his seminal work in so many areas provided the basis for further

Ivan Vasilievich Vernadsky with his family. Vladimir is standing on the far right.

research in entirely new fields of research: genetic mineralogy, geochemistry, hydrogeochemistry and hydrogeothermy, oceanography, radiogeology, cryology or the study of permafrost, and cosmochemistry. He virtually created the field of biogeochemistry and his insistence on studying the chemistry of other planets to find the similarities—and dissimilarities—to our own, foreshadowed much of the work that would reach fruition after his death in the manned space program.

In all these areas Vernadsky left his imprint. And in his extensive work as a teacher and scientist he also left an extensive school of scientific thought that still makes itself felt in Russia today. In fact, one might say with justification, that Russian science is still on the "cutting

Russian Academy of Sciences

Vladimir Vernadsky with other members of the Russian Duma circa 1905.

the Kadets from 1903 until 1917 and for brief periods in the Duma as a Kadet delegate.

When the Bolsheviks took power, Vernadsky, diagnosed with tuberculosis, removed himself to his country home in Ukraine. While in Ukraine in 1919-1920 he set up the Ukrainian Academy of Sciences and established in the capital, Kiev, the National Library of Ukraine which still bears his name. When Kiev fell to the Bolsheviks, Vernadsky withdrew to Crimea, still under the control of the Whites. Here he was elected president of the Tauride University.

When Crimea fell to the Bolsheviks, Vernadsky was considering emigrating to the United States where he hoped he would be able to set up a Biogeochemical Laboratory under the Carnegie Institute. But his election to the presidency of Tauride University and a deep-rooted concern for the fate of Russian science under Bolshevik rule, kept him in Crimea where many Russian intellectuals had sought refuge. With the fall of Crimea to the Bolsheviks, Vernadsky, although known as an active member in the Kadet Party, was brought back to St. Petersburg, not as a prisoner, but in order to again take up his position at the

edge" largely thanks to the "Vernadsky school," which, of course, would include not only his own students, but theirs as well, as well as the numerous individuals who have been attracted to science by the work and example of Vernadsky.

While he worked half his life in Tsarist Russia and the other half under the Soviet regime, he was an adherent of neither. His devotion was to the nation, and he was democratic in spirit, putting him somewhat at odds with both of these political systems. In his younger days, prior to the Bolshevik Revolution, he had been extremely political. His father, Ivan Vasilievich Vernadsky, was a prominent Russian economist who introduced the work of American economist Henry Charles Carey to Russian circles and helped lay the basis for the great reforms of the 1860s. Vladimir was deeply involved in the reform movement of his own time, helping to transform the illegal Union of Liberation (which he helped to establish), into the Constitutional Democratic Party (Kadets) when political parties were finally permitted in Russia after the 1905 Revolution. Vernadsky served on the Central Committee of

Russian Academy of Sciences

Teaching geochemistry at the Higher Women's Course in St. Petersburg.

Florentine physician and scientist Francesco Redi (1627-1697).

Mineralogical Museum which he had left three years before. Lenin's policy of broad electrification of the Soviet Union necessitated a revival of the old scientific cadre from the pre-war period. Vernadsky, who had been a teacher and a mentor for Lenin's brother, Alexander, prior to Alexander's involvement in an attempted assassination of the Tsar in 1881, was also not totally unknown to Lenin.

During the often tumultuous and difficult years following the Russian Revolution and civil war, Vernadsky would steadily work to revive and advance Russian science. Until the mid 1930s, he was permitted to travel abroad almost every year, consolidating contacts with the main figures in international science, with Marie Curie in Paris, with Otto Hahn in Germany, and with Lord Rutherford and Frederick Soddy in England.

Vernadsky almost single-handedly conducted a campaign in Russia to establish a major research center for nuclear energy. Already in 1921 he had succeeded in creating the *Radium Institute in St. Petersburg,* but the Soviet leadership was slow to realize the importance of this research. At the beginning of World War II when Vernadsky began to suspect work on the atom in the U.S. and elsewhere for military reasons, he insisted that the Russian Government move quickly on the matter, and was initially the organizer of the effort. As the program moved closer to weapons development, Vernadsky was effectively cut out of the program, the authorities viewing the ageing scientist as still something of a dissident and therefore not entirely trustworthy.

Indeed, although a patriotic Russian even in Soviet times, Vernadsky never accepted the tenets of dialectical materialism. As the Bolshevik regime in the late 1920s attempted to take over the Academy of Sciences and bring the old "gray beards" under strict supervision by the orthodox Marxists, Vernadsky led the fight to maintain the independence—and the intellectual integrity—of the Academy and the Academi-

cians. Needless to say, he was only partially successful. While the years following 1928 would see an influx of academics from the Party hierarchy into the Academy, Vernadsky attempted to work with those who were intellectually qualified and to limit the damage inflicted on the Academy by those who were not.

And although Vernadsky was barely tolerated by the Party apparatchiks, accused of being a "vitalist" because of his views on the question of life, he was also "protected," by higher authority from the machinations of the NKVD (the predecessor to the KGB) because of his intellectual preeminence, and continued to exert something of an influence on the scientific elites. He utilized his rather unique position to try to save

Louis Pasteur discovered chirality in living cells.

many of his colleagues from being sent to the Gulag, or, if sent, to get them into a situation in which they could continue doing some form of useful scientific work, and the possibility for such work even in the Gulag became greater after World War II began. A year before he died, Vernadsky was awarded the Hero of Socialist Labor. Half of the money connected with the award, Vernadsky donated to the war effort.

But Vernadsky is most noted for his work on the biosphere and the question of life in the universe. From the beginning he refused to accept the basic premise of abiogenesis, the idea that life proceeded from a combination of inorganic materials, oxygen, carbon, nitrogen which combined in some mysterious way, to become living matter. Vernadsky saw no scientific evidence that such a process ever occurred. He adhered to the principle enunciated by the 16th Century Italian physician, Francesco Redi, *omnium vivum e vivo,* that life only proceeds from life. This was also the conclusion from the 19th Century work of Louis Pasteur, who discovered the notion of chirality or right- or left-handedness in living tissue. This indicated that living tissue had a decidedly different structure than inorganic matter, giving more scientific grounding to the thesis of Redi.

Vernadsky was convinced

The Radium Institute in St. Petersburg.

A Russian delegation visits with scientific colleagues in Berlin, 1928. Vernadsky on the far right and Albert Einstein, third from the left.

Vernadsky also was the first to recognize the absolutely essential role of the biosphere, i.e. the total aggregate of living matter on Earth, in the development of the Earth's upper crust and atmosphere and stratosphere. With the appearance of Man, however, Vernadsky saw an entirely new dimension in the history of the biosphere in the changes wrought through the productive activity of Man. Just as the biosphere is characterized by a steady increase in its energy throughput as it develops and subsumes the Earth, so also does the activity of Man begin to develop its own characteristic form of "energy" which assumes a predominant role in the biosphere and transforms it.

that there was no indication within geological time (which we can examine through a study of the Earth's crust), of life ever proceeding from non-life. He was also convinced that we would not find indications of abiogenesis in cosmic time either, that is, during the earlier period when Earth was forming out of its swirling vortex, although this latter era was more difficult to investigate. Secondly, given the continual exchange of matter between our Earth and the surrounding space, in the form, for instance, of cosmic radiation or cosmic particles, Vernadsky noted, life may well have been brought to us from elsewhere and, finding ideal conditions here, developed and flourished in that environment. Vernadsky urged the examination of material from other planets, such as meteorites, in order to determine their chemical composition, and possibly, if there were also there signs of life. Vernadsky held to his thesis despite the consistent attempts by orthodox Marxist scholars, who deemed Vernadsky's attacks on the theory of abiogenis undermining the "materialistic" foundations of their own "dialectical materialism," to disprove it. The career of science "apparatchik," Alexander Oparin was carefully cultivated by Vernadsky's enemies in order to discredit Vernadsky's hated "vitalism." The "fellow traveler" networks of Bertrand Russell and J.B.S. Haldane helped to make Oparin's 1936 book *The Origin of Life* the bible of the abiogensists. Oparin was feted by these Western circles as a great scientific thinker in spite of the key and very public role he played in the Soviet Union in promoting the frauds of that notorious fraud, Trofim Lysenko, who led a campaign to eliminate some of the most important scientists in the Soviet Union.

Vernadsky called this new era with the development of man, the noösphere, after the Greek term noos (or mind), to distinguish it from the biosphere per se. The term was coined by Eduard LeRoy, who, together with Jesuit palaeontologist, Teilhard de Chardin, attended Vernadsky's geochemistry lectures in Paris at the Sorbonne in 1924. Vernadsky adopted the term as his own to depict the stage of the biosphere characterized by the preponderant activity of man.

Vernadsky felt that now in the 20th Century, with Einstein's discovery of relativity and with the mastery of atomic energy, man was in the process of taking a tremendous leap forward in the development of the noösphere, putting him on the verge of extending his reach into the surrounding universe. His last great works, the unfinished "The Chemical Structure of the Biosphere and Its Environs" and "Scientific Thought As A Planetary Phenomenon" both written between 1931 and 1944, were to be the final word of his mature thought. Lamentably, the first work, more broadranging than the latter, was to have a third section devoted exclusively to the notion of the noösphere, but Vernadsky was not able to conclude the work before his death. Given that critical lacuna, the second work, "Scientific Thought As a Planetary Phenomenon" from which this chapter is taken, undoubtedly represents Vernadsky's most extensive elaboration of the notion of the noösphere.

The chapter appears in the section of the book entitled

New Scientific Knowledge and the Transition from the Biosphere to the Noösphere. In it, Vernadsky traces the development of man from his first appearance as man with his mastery of fire, the first instance that we are aware of, in which man takes direct control of a force of nature. Vernadsky indicates here also the new possibilities for man's role in the universe, the possibility of extending his activity into space and possibly to other planets. It is imbued with a tremendous sense of optimism, optimism which, by the way, never abated, even in the face of the horrors of World War II.

Quite simply, Vernadsky understood that there existed in the universe a principle of development, which, with the development of man and the new-found role of man's reason, expressed itself in the necessity for continued progress. While a great deal of distortion of the thrust of Vernadsky's thought has been introduced into the public domain over the last several decades by the Green movement's "adoption" of Vernadsky as some form of "ecologist," it is hoped that the ideas expressed clearly by Vernadsky in the present work will lay to rest any doubts about where he stood in that respect, firmly behind the commitment to the scientific and technological development by means of which man becomes ever more the master of his universe.

Russian Academy of Sciences

Vernadsky in his study around the time of the writing of "Scientific Thought As A Planetary Phenomenon."

The Transition From the Biosphere To the Noösphere

by Vladimir Vernadsky

Excerpt from "Scientific Thought As a Planetary Phenomenon"

Chapter VII

100.

The sciences concerned with the biosphere and its objects, that is, all of the humanities without exception, the natural sciences in the proper sense of the term (botany, zoology, geology, mineralogy, etc.), all the technical sciences, — applied sciences broadly understood — appear as areas of knowledge, which are the most accessible to the scientific thought of Man. Here we have concentrated millions upon millions of continuous scientifically established and systematized facts, which are the result of organized scientific labor, and which inexorably increase with each new generation, rapidly and consciously, since the 15th to 17th centuries.

In particular, the scientific disciplines dealing with the structure of the instruments of scientific cognition, indissolubly linked to the biosphere, may be scientifically viewed as a geological factor, a manifestation of the manner in which the biosphere is organized. These are sciences dealing with the "spiritual" creativity of the human individual in his social environment, the sciences of the brain and of the sense organs, of the problems of psychology or logic. These condition the quest for the fundamental laws of Man's scientific cognition, that is, those powers which have transformed the biosphere encompassed by Man into a natural body, new in its geological and biological processes, into a new state, the noösphere,[1] consideration of which I will turn to below.

Its creation, beginning intensively (in the measure of historical time) some tens of thousands of years ago, was an occurrence of extreme importance in our planet's history, connected above all with the growth of the science of the biosphere, and was definitely not by chance.[2]

We may therefore state that the biosphere represents the fundamental sphere of scientific knowledge, although only now are we on the point of distinguishing it from its surrounding reality.

101.

It is clear from the foregoing that the biosphere is equivalent to "nature" in the ordinary sense of the term, as this term is used in the deliberations of the naturalist and in philosophical discussions, where it does not refer to the Cosmos at large but rather to phenomena contained within the bounds of Earth. In particular, it corresponds to the naturalist's nature.

Not only is this "nature" not amorphous and without form, as was thought for centuries, but rather it possesses a determined, well-defined structure,[3] which, as such, must be reflected and taken into consideration in all the conclusions and deductions relating to nature.

In scientific investigations it is especially important not to forget this and to examine it, since unconsciously, sci-

---

1. *E. Le Roy. Les origines humaines et l'évolution de l'intelligence, III. La noosphère et l'hominisation.* Paris, 1928, pp. 37-57

2. I will return later to that process. Here I will merely indicate the thought of Le Roy: "Deux grands faits, devant lesquels tous les autres semblent presque s'évanoir, dominent donc l'histoire passée de la Terre: la vitalisation de la matière, puis l'hominisation de la vie." — op. cit. p. 47

3. That "structure" is quite distinctive. It is not a mechanism nor is it something stationary. It is dynamic, ever changing, mobile, in each instance changing itself and never returning to a previous form of equilibrium. The closest to it is the living organism, distinguishing itself, however, from it in its physical-geometrical state of its space. The space of the biosphere is heterogeneous in its physical-geometrical dimension. I think, that it is proper to assign to that structure a specific notion of organization.

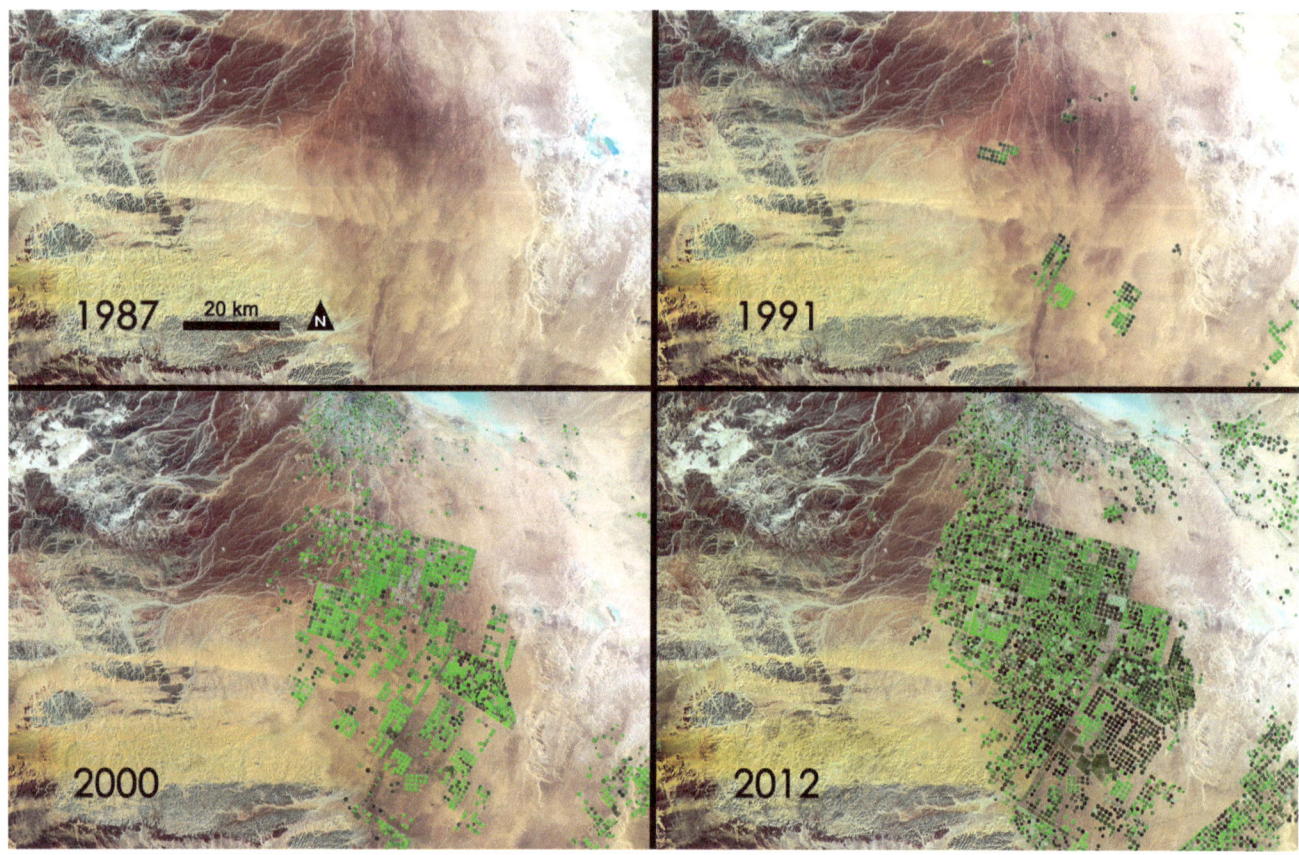

NASA

The energy of human culture: The greening of the desert near the city of Tubarjal in Saudi Arabia.

entists and scholars, when contrasting the human individual with nature, are overwhelmed by the grandeur of nature against the human individual.

But life in all its manifestations, including the activity of the human individual, radically transforms the biosphere to the degree that not only the aggregate of indivisible life, but even some problems of the solitary individual person in the noösphere, cannot remain without consideration in the biosphere.

102.

Living nature is the fundamental trait of the manifestation of the biosphere, and by this clearly distinguishes itself from the Earth's other envelopes. The structure of the biosphere is characterized first and foremost by life.

We see below that there lies, in a number of aspects, an unbridgeable gulf between the physical-geometric properties with regard to the weight and quantity of the atoms in living organisms—in the biosphere they are manifested in the form of their aggregates-living substance, and such properties, in inert matter, which comprises the overwhelming part of the biosphere. Living matter is the bearer for, and creator of, free energy, not

existing to such a degree in any one of Earth's envelopes. This free energy—biogeochemical energy[4]—embraces the entire biosphere and fundamentally determines its entire history. It stimulates and radically transforms the intensity of the migration of the chemical elements which compose the biosphere and determines its geological significance.

4. The concept of biogeochemical energy came to me in 1925 in a still unpublished paper for the L. Rosenthal Fund (the fund is no longer in existence). This fund gave me the opportunity to quietly devote myself to this work over the course of two years. A series of articles and books from this research are therefore in print:
• Biosfera. Leningrad, 1926, pp.30-48;
• Etudes biogéochimiques, 1. Sur la vitesse de la transmission de la vie dans la biosphère.—*Isvestiia* AH, 6 series, 1926, v. 20, No. 9, pp. 727-744;
• Etudes biogéochimiques. 2 La vitesse maximum de la transmission de la vie dans la biosphère.—*Izvestiia* ANs, series 6, 1927, V. 21, No. 3-4, pp. 241-254;
• O razmnozhenii organizmov i ego znachenii v mekhanizme biosfery. *Izvestiia* AN, series 6, 1926, V. 20, No. 9, pp. 697-726, No. 12, pp.1053-1060;
• Sur la multiplication des organismes et son role dans le mécanisme de la biosphère, pp. 1-2, *Revue générale des sciences pures et appliquées.* Paris, 1926, t. 37, N 23, pp. 661-668; pp. 700-708;
• Bakteriofag I skorost' peredachi zhizni v biosfere, *Priroda,* 1927, No. 6, pp. 433-446.

During the past ten thousand years, a new form of this energy has been created within the realm of living substance, even more intense and complex, and rapidly growing in importance. This new form of energy, associated with the vital activities of human societies, of the genus *Homo* and other closely related genera (hominids), while preserving the expression of ordinary biogeochemical energy, brings about simultaneously new forms of migration of chemical elements, which in their diversity and power leave the ordinary biogeochemical energy of the living matter of the planet far behind.

This new form of biogeochemical energy, which might be called the energy of human culture or cultural biogeochemical energy, is that form of biogeochemical energy, which creates at the present time the noösphere. Later I will return to a more detailed exposition and analysis of our understanding of the noösphere. But at the moment it is only necessary for me to present a brief outline of its manifestation on our planet.

This form of biogeochemical energy is proper not only to *Homo sapiens*, but to all living organisms.[5] However, among these, this energy appears insignificant compared with ordinary biogeochemical energy, and is barely noticeable in the balance of nature, and then only on the scale of geological time. It is associated with the mental activity of organisms, with the development of the brain in higher forms of life, and only with the appearance of reason do its effects produce the form of transition of the biosphere into the noösphere.

Its manifestation in the predecessors of Man was probably developed over the course of hundreds of millions of years, but it was able to express itself as a geological force only in our time, when *Homo sapiens* has embraced the entire biosphere with his life and cultural work.

103.

The biogeochemical energy of living matter is determined primarily by the reproduction of organisms, by their unremitting endeavor (determined by the energetics of the planet) to achieve a minimum of free energy — determined by the fundamental laws of thermodynamics corresponding to the existence and stability of the planet.

It is expressed in the respiration and alimentation of living organisms, by "the laws of nature," which to the present time had not found a mathematical expression, although the task of discovering such was clearly posed already in 1782 by Christian Wolff at the former St. Petersburg Academy of Sciences.

Certainly, this form of biogeochemical energy is also characteristic of *Homo sapiens*. For Man it is, as for all other organisms, a "species characteristic,"[6] and seems to us invariable in the course of historical time. In other organisms, there is another form of "cultural" biogeochemical energy, which is unchangeable or only slightly so. This other form is manifested in the everyday life or technical conditions of life of the organisms—in their movements, in their daily existence and the construction of their habitats, in their displacement of other organisms in their environment, etc. As I have already noted, this energy makes up only an insignificant part of their biogeochemical energy.

With Man, however, the form of biogeochemical energy connected to reason grows and expands with time, rapidly moving to the fore. This increase is possibly related to the growth of reason itself—a process which seems to occur very slowly (if at all) but is chiefly connected to its honing and deepening in using it to consciously transform the social environment, and is especially due to the growth of scientific knowledge.

I shall proceed from the fact that in the course of hundreds of thousands of years, *Homo sapiens* skeletons, including the craniums, do not provide a basis for considering them as belonging to a different species of Man. This is assumed only on the condition that the brain of Paleolithic Man was not in some fundamental way structurally distinct from the brain of contemporary Man. At the same time, there can be no doubt that the mind of man during the Paleolithic period for that particular species of *Homo* cannot bear comparison with the mind of modern man. Hence it follows that reason is a complex social structure, erected similarly for contemporary Man, as well as for Paleolithic Man, on the same neural substrate, but in different social circumstances that formed over time (essentially over space-time).

Its explicit transformation is a fundamental element leading ultimately to the transformation of the biosphere into the noösphere, first and foremost, through the creation and growth of the scientific understanding of our surroundings.

104.

The creation on our planet of cultural biogeochemical energy appears to be a fundamental fact of its geological history. The way was prepared in the course of all geological time. The fundamental determining process here is the maximum expression of the human mind. But in essence this is inextricably linked with the totality of all biogeochemical energy of living matter.

By means of the migration of atoms in living processes, life bundles together into a single whole all the migrations

5. V. I. Vernadskii. *Biosfera*, pp. 30-48; O razmnozhenii organizmov i ego znachenii v mekhanizme biosfera.—op. cit., No. 9, pp. 697-726, pp.1053-1060.

6. Regarding species characteristics, See: V. I. Vernadskii. Considerations générales sur l'étude de la composition chimique de la matière vivante.—*Trudy Biogeokhimicheskoi laboratorii*, T. 1, 1930, pp. 5-32.

of atoms in the non-living matter of the biosphere.

Organisms are alive only until the material and energetic exchange between them and the surrounding biosphere ceases.[7] In the biosphere certain grand chemical circulatory processes of atomic migrations appear, in which living organisms are involved, as a lawful inseparable, and often fundamental, part of the process. These processes remain unchanged in the course of geological time: for instance, the migration of the atoms of magnesium forming into chlorophyll has gone on uninterruptedly for at a least two billion years through countless genetically interconnected generations of green organisms. Living organisms, continuously and inseparably connected to the biosphere through such migrations of atoms alone, constitute a lawful part of its structure.

This must never be forgotten in our scientific study of life and in our scientific judgments regarding life's manifestations in Nature. We must not neglect to take into account that this indissoluble material and energetic link of living organisms with the biosphere—a link of a completely distinct character, which is "geologically eternal," and may be expressed with scientific accuracy—is always present in any of our scientific approaches to living things and must be reflected in all of our logical conclusions and deductions concerning them.

Coming to the study of the geochemistry of the biosphere, we must above all precisely estimate the logical importance of that connection, which must necessarily enter into all of our constructs regarding life. It is independent of our will and cannot be excluded from our experiments and observations; it must always be taken into account as something fundamental that is inherent in living things.

In this way the biosphere without exception must be reflected in all of our scientific judgments. It must be manifest in every scientific experiment and observation—and in all a human individual's deliberations, and in all speculation, from which a human individual—even in his thoughts—cannot refrain.

The mind can therefore be manifest to the maximum degree only under conditions of the maximum development of the fundamental form of human biogeochemical energy, that is, under the condition of man's maximum degree of reproduction.

105.

The potential possibility of the expansion over the surface of the entire planet by the multiplication of a single organism, of a single species, is proper to all species,

since, for all of them the law of reproduction is expressed in one and the same form, in the form of a geometrical progression. I have long emphasized the fundamental significance of this phenomenon for biogeochemistry,[8] and I shall return to it in its proper place in this book.

It is evident that the phenomenon of the expansion over the entire surface of the planet by a single species developed widely in the case of aquatic life such as microscopic plankton in lakes and rivers, and some forms of microbes, essentially also aquatic, on the cover of the planet's surface, and disseminated through the troposphere. For larger organisms, we observe this almost in full measure with certain plants.

For Man this begins to be seen in our time. By the 20th Century the entire globe and all the seas have been encompassed by Man. Owing to the progress of communications, mankind is able to be in continual contact with the entire world; nowhere solitary or helplessly lost in the immensity of Earth's nature.

Now the Earth's human population has reached the previously unprecedented figure of nearly two billion people, despite the losses incurred by wars, starvation and disease, which continuously afflict hundreds of millions of people, and which have seriously retarded the course of that process. It will require, however, only an insignificant amount of time in geological terms, barely more than a few hundred years, for such relics of barbarism to cease. This could of course be accomplished even now: the possibility lies already within Man's grasp, and a will informed by reason will inevitably embark on this path, since it corresponds to the natural thrust of the geological process. This is even more the case as the opportunities for doing this are rapidly, almost spontaneously, increasing. The real significance of the popular masses, which have endured these sufferings more than anyone, is irrepressibly growing.

The number of people inhabiting our planet began to increase approximately 15,000 to 20,000 years ago, when Man became less constrained by the shortage of food with the development of agriculture. Presumably at that time, around 10,000-18,000 years ago, the first leap in man's reproduction took place.[9] G.F. Nicolai (in 1918-1919)[10] attempted to quantitatively determine the actual multiplication of Man and the development of agriculture, that is, Man's real colonization of the planet. In his

---

7. The complete absence of exchanges in dormant forms of life can not yet be considered verified. It is extremely slow—and perhaps in some cases of atomic migrations actually absent—this becomes noticeable only in geological time..

8. Regarding the speed of the transmission of life, see: V. I.. Vernadskii. *Biosfera*, p. 37-38; Also Etudes biogéochimiques. 1. Sur la vitesse de la transmission de la vie dans la biosphere.—op.cit., N 9, pp 727-744; Also *Biogeokhimicheskie ocherki*, (1922-1932) M.-L. 1940, pp. 59-83.

9. V.G. Childe. *Man Makes Himself.* London, 1937, pp. 78-79.

10. G.F. Nikolai. *Die Biologie des Krieges.* 1.—Betrachtungen eines naturforschers den Deutschen zur Besinnung. Band 1. Zurich, 1919. p. 54.

Stefan Kühn

A new form of energy connected to the atomic nucleus. Nuclear reactors in Cattenom , France.

calculations, which encompassed the entire land-mass of the globe, there were 11.4 people per square kilometer, which constituted 2.10% to 4% of the possible colonization. Taking into account the energy received from the Sun, agriculture makes it possible to support 150 people on one square kilometer, that is, on the entire land area of the globe you could support a population of 22.5 billion individuals, that is, 22-24 times the number that now inhabit it.[11] But Man obtains energy for nourishment and subsistence not only from agricultural labor. Taking into consideration that possibility, Nicolai made a rough estimate that, starting from the historic epoch begun in our time, utilizing new energy sources, Earth could support a population of 3 trillion people, that is, more than ten million times greater than the present population. Now after more than 20 years have passed since Nicolai made his calculations, these figures ought to be greatly increased since Man may now use a source of energy to which Nicolai in 1917-1919, gave no thought: namely, the energy connected to the atomic nucleus. We should now more simply say that the source of energy subject to Man's reason, in this energetic epoch of the life of Man on which we have now embarked, is practically unlimited. From this it is also clear that cultural biogeochemical energy possesses the same quality. In the calculations of Nicolai in his day, machines increased the energy of Man more than tenfold. We cannot now give a more precise figure, yet recent estimates by the American Geological

Committee indicate that hydroelectric power, utilized now over the entire globe, had reached, by the end of 1936, the level of 60 million horsepower: Within 16 years it had increased by 160 percent, chiefly in North America.[12] As a result, we must increase the calculations of Nicolai by more than one and a half times.

Actually, all of those calculations of the future, expressed in numbers, are not significant, since our knowledge of the energy available to mankind may be said to be rudimentary. Certainly the energy available to Man is not an unlimited quantity, as it is limited by the dimensions of the biosphere. These also define the limits of cultural biogeochemical energy.

We will see that there is even a limit to the fundamental biogeochemical energy of mankind-namely, the speed of the transmission of life, the limit of the reproduction of Man.

The speed of colonization, the quantity V, essentially considered by Nicolai, is based on actual observations of the colonization of the planet by Man under clearly inauspicious conditions of life. In addition, we will furthermore see that there exist some phenomena in the biosphere yet unknown to us, but powerful in a given geological era and under certain conditions of the ecosystem, which lead to a stationary maximum number of individuals per hectare.

106.

It was only at the beginning of the 19th Century that we were able to determine with any accuracy the number of human beings living on the planet. The number was arrived at with a great degree of possible error. In the last 137 years, our knowledge has increased considerably, but it still does not achieve the accuracy required by modern science. For earlier periods, the figures are only provisory. All of this aids us, however, in understanding the underlying process.

Regarding this, the following facts may be of some significance.

The number of people in the Paleolithic period probably reached a few million. Presumably this developed from one single branch. But the opposite may also be true.[13]

During the Neolithic period, there were probably some tens of millions of people on the entire surface of the

11. G.F. Nikolai. op.cit., p. 60.

12. Water-Power of the World (News and Views).—*Nature*, 1938, v. 141, N 3557, p. 31.

13. See Le Roy. op.cit.

NASA

The formation of phyloplankton off the coast of New Zealand, Oct.2009

Earth. It is possible to assume that even in historical time, the population did not reach 100 million, or perhaps slightly more. [14]

For 1919, G.F. Nicolai surmised that the population of the planet increased annually by 12 million people, that is, a daily increase of approximately 30,000 people. According to the critical report of Kulischer (1932),[15] world population in 1800 reached 850 million people (A. Fisher gives an estimate equal to 775 million). We can estimate the population of the white race in the year 1000 A.D. as being equal to 30 million in all, in 1800, 210 million, and in 1915, 645 million. The entire population in 1900, according to Kulischer, was around 1,700 million, but according to A. Hettner (1929),[16] the number was 1,564 million in 1900 and 1,856 in 1925.

Evidently in our own time this number has reached around 2 billion people, more or less. The population of our own country (around 160 million people) makes up around 8% of the total world population. The total population is rapidly increasing and, apparently, the percentage of our country's population is increasing relatively, as

its increase is greater than that of the world average. In general, we should expect by the end of the century a significant increase exceeding 2 billion people.

107.

The multiplication of organisms, that is, the manifestation of biogeochemical energy of the first type without which there is no life, is inseparable from Man. But Man, from the very moment that he distinguishes himself from the aggregate of other life-forms on the planet, already possesses the tools, albeit rather primitive ones, which allow him to increase his muscle-power and is the first expression of contemporary machinery which distinguishes him from other living organisms. The energy which sustained him was, however, produced through the alimentation and respiration of his own organism. It is likely that already for hundreds of thousands of years as Man— the genus *Homo,* and his predecessors—he possessed tools made out of wood, bone and stone. Slowly, in the course of many generations, he developed the ability to fashion and utilize those tools, honing that capability, reason in its initial manifestation.

Those tools had been observed already in the earliest Paleolithic period, 250-500,000 years ago.

In that period, a significant part of the biosphere experienced a critical time. It seems that already at the end of the Pliocene period, abrupt changes occurred—in the water and heat regime of the biosphere, beginning and continually developing during the period of glaciation. We are apparently still living in the period of the last gla-

14. B.P. Weinberg. *Twenty thousand years from the beginning of the elimination of the oceans. Review of the History of Mankind from a Primitive State to the Year 22,300. (A scientific fantasy).* Sibirskaya priroda. Omsk, 1922, No. 2 p. 21 (assumes a population of 80 million at the beginning of our era).

15. A. and E. Kulisher. *Kriegs-und Wanderzüge. Weltgeschichte als Völkerbewegung.* Berlin—Leipzig, 1932. p. 135

16. A. Hettner. *Der Gang der Kultur Uber die Erde.* 2nd edition, revised and expanded. Aufl. Leipzig-Berlin, 1929. p. 196

ciation's retreat phase, although we don't know whether this is permanent or merely temporary. In that half million years, we see sudden fluctuations in the climate; relatively warm periods—lasting tens and hundreds of thousands of years—gave way in the northern and southern hemisphere to periods, when large masses of ice slowly moved (measured in historical time), reaching the thickness of a kilometer, for instance, in the vicinity of Moscow. These disappeared from the Leningrad region 7000 years ago,[17] but still envelop Greenland and Antarctica. Apparently, *Homo sapiens* or his closest predecessors appeared not long before the onset of that glacial period, or in one of its warmer episodes. Man survived the severe cold of that period, possibly due to the great discovery that had been made in the Paleolithic age— the mastery of fire.

Man's discovery of fire was the first sign of the noosphere

That discovery was made in one, two, or possibly more places, and slowly spread among the peoples of the Earth. It seems that we are dealing here with a general process of great discoveries, in which it is not the mass action of mankind, smoothing and refining the details, but rather the expression of separate human individuals. As we'll later see, we can track this phenomenon more closely in very many cases nearer to our own era.

The discovery of fire presents the first instance in which a living organism takes possession of, and masters, one of the forces of nature.[18]

Undoubtedly this discovery lies, as we now see, at the foundation of mankind's subsequent future increase and of our present powers.

But that increase occurred extraordinarily slowly, and it is difficult for us to imagine the conditions under which it may have occurred. Fire was already known to the ancestors or the predecessors of that species of hominid, which established the noösphere. The recent discovery in China reveals to us the cultural remains of Sinanthropus, which indicate an extensive utilization by him of fire, apparently long before the last glacial period in Europe, hundreds of thousands of years before our time. We have at present no reliable data as to how that discovery was made. Sinanthropus already possessed reason, had crude tools, used speech, and conducted funereal rites. He was already Man, but was distinguished from us by a number of morphological characteristics. We don't exclude the possibility that this was one of the ancestors of the present population of China. [19]

108.

The discovery of fire is all the more remarkable in that the appearance of fire and light in the biosphere before Man was a relatively rare phenomenon and generally occurred when it encompassed a large space, as in forms of "cold light," which are expressed in heavenly luminescence, polar light, silent electric discharges, stars and planets, or luminescent clouds. But only the Sun, that source of life, brightly displays simultaneously both light and heat, illuminating as well as warming the dark planet.

Living organisms for a long time produced a form of "cold light." This was seen in such imposing phenomena as bioluminescent oceans, which encompass areas usually stretching hundreds of thousands of square kilometers, or in the luminescence of the ocean's depths, the significance of which is only now beginning to be understood. Fire, accompanied by high temperatures, was manifest in local phenomena, rarely encompassing a vast expanse as, for example, in volcanic eruptions.

But these phenomena, grand by human standards, obviously owing to their great destructive force, in no way assisted in Man's discovery of fire. Man had to have sought it in natural phenomena closer to him, and in less unusual and dangerous forms than volcanic eruptions, which even now exceed in their magnitude, the powers of contemporary Man. We are only now beginning to achieve its practical utilization in circumstances far beyond the power or the imagination of Paleolithic Man.[20]

---

17. Now we know that in the environs of Leningrad, the ice disappeared about 12 thousand years ago. (Ed.)

18. V.G. Childe. *Man Makes Himself*. London, 1937, p. 56. Compare: J.G. Frazer. *Myths of the Origin of Fire*. London, 1930.

19. See: On the technology of Sinanthropus and the use of fire by him, see: B.L. Bogalevskii. The technology of primitive-communistic societies.—*Istoriia tekhniki*, vol. 1, ch. 1, Moscow-Leningrad, 1936, pp. 26-27. Fire was also used by Pithecanthropus which lived earlier in the very beginnings of the Pleistocene, scarcely more than 550 thousand years ago. Compare B. L Bogalevskii, Ukaz. soch., pp. 11, 67. The use of fire for Pithecanthropus is still not proven, but is highly probable.

20. Only in the 20th Century in Larderello on the initiative of Le Conte, did Man obtain, with the help of drilling, superheated vapor (140 degrees C) as a source of energy. Still later in Soffioni (New Mexico) and Sonoma that method was improved. Before his death, Parsons was working on a practical design to achieve, by drilling, an inexhaustible source of energy, at least from the point of view of humanity, from the internal heat of the Earth's crust. An analogous attempt of obtaining energy from the cold depths of the ocean by French academic Claude did not succeed in doing so only because of some acts of criminal hooliganism in 1936. We doubtless have in the hands of Man in these developments a practically inexhaustible force.

He would have had to have sought sources of heat and fire in surrounding everyday phenomena; in the places where he lived—in the forests, on the steppes, in the midst of a living nature with which he lived in close intercourse (now long forgotten). Here he would encounter fire and heat in non-threatening forms in a succession of commonplace phenomena. These were, on the one hand, fires in which living matter, living and dead, was burned. These were precisely those sources of fire used by Paleolithic Man.

He burned trees, plants, bones, the very same that fed the fires around him, independent of his will. Until Man began to use it, fire was caused by two very different sources. On the one hand, lightning discharges caused forest fires or ignited dry grass. Even now Man suffers from such fires. The conditions of nature during the glacial period, particularly during the interglacial period, may have provided even more favorable conditions for such thunderstorm phenomena. Yet there was also another source for the independent occurrence of fire.

These were the life-activities of the lower organisms which led to the fires in the dry steppes,[21] to the burning of layers of coal deposits, to the burning in peat bogs, which endured for several generations and provided convenient opportunities for obtaining fire. We have direct evidence of such coal fires in the Altai region, in the Kuznetz basin, where they occurred in the Pliocene and in the post-Pliocene period, but where they continued into the historical period, and where they must be considered still occurring now. The causes of these fires has to this day not been fully clarified, but everything indicates that we here are hardly dealing with the result of a purely chemical process of spontaneous combustion, that is, one caused by the intensive oxidation of fractured coal by oxygen in the air, or by the spontaneous combustion of the sulphuric compounds of iron produced by the heat developing during oxidation of the coal.[22]

More probable is the existence of biochemical processes connected to the life-activities of thermophilic bacteria. Regarding peat bogs, we even have during the recent period the direct observations of B.L. Isachenko and N. I. Malchevskoy.[23]

These phenomena now require more careful study.

109.

Such warm regions, winter and summer, as well as areas with hot springs, were blessings of nature for Paleolithic Man, who also had to utilize them as they were, or have recently been, utilized by tribes or peoples that we still find living in a Paleolithic state.

With the great powers of observation of Man in that era and with his close proximity to nature, such areas undoubtedly attracted his attention, and would have been utilized by him, particularly in the glacial period.

It is interesting that among the instincts of animals we can observe the use of those same biochemical processes. This is seen in the family of cocks, the so-called brush turkeys, or the big-nosed megapodes of Oceania and Australia, which utilize the heat of fermentation, that is, a bacterial process, for hatching the fledging out of the egg, building large piles of sand or earth and mixing it with strongly rotting organic remains.[24] Those piles can reach 4 meters in height and a temperature of not under 44 degrees Centigrade. These seem to be the only birds possessing such an instinct.

It is possible that ants and termites purposefully raise the temperature of their dwellings.

But these feeble endeavors cannot be compared with that planetary revolution produced by Man.

Man utilizes as a source of energy, fire, the products of life—dried vegetation. Many myths about its discovery have been coined and kept in circulation.[25] But most typical is that Man utilized methods for creating it which would never have been observed by him in the biosphere until his discovery of it. The most ancient technique seems to have been the transfer of muscle-power into heat (powerfully rubbing together dry objects), and creating sparks, and catching them from certain rocks. The complex system of maintaining fire after all came into ex-

21. Some people deny that spontaneous combustion of dry grass in the steppes, pampas, and forests actually occurs. Nowadays fires are nearly always caused by Man, but there are occasions which, it seems, indicate without a doubt the possibility of a process of spontaneous combustion in the steppes as a result of the direct activity of the Sun. The causes of the phenomenon are not clear. Concerning such events, see: E. Poepping. *Reise in Chili, Peru und auf dem Amazonenstrom wahrend der Jahre 1827-1832*, Bd. 1. Leipzig, 1835, p. 398. G.D. Hale Carpenter. *A Naturalist on Lake Victoria.—With an Account of Sleeping Sickness and the Tse-tse Fly.* London, 1920, pp. 76-77.

22. See: M. A. Usov. *Composition and tectonics of the deposits of the Southern region of the Kuznetsk coal basin.* Novonikolaevsk, 1924, p. 58; idem, Subterranean first in the Prokpyevsk region—a geological process, *Vestnik Zapadno-sibirskogo geologo-razvedochnogo trests*, 1933, no. 4, p. 34; and V.A. Obruchev. Subterranean fires in the Kuznetsk basin, Priroda, 1934, no. 3, pp. 83-85. Already I.F. Hermann, who discovered the Kuznetz coal basin, indicated in 1796 these phenomena. See: B.F. J. Hermann. "Notice sur les charbons de terre dans les environs de Kouznetz en Siberie." *Nova acta Academiae scientuarum Imperialie Petropolitanae.* St. Petersburg, 1793, pp. 376-381. Compare: V. Yavorsky and L.K. Radugina. Die Erd-

braende im Kuznetzk Becken und die mit ihnen verbundenen Erscheinungen, *Geologische rundschau.* 1933, Hf. 5; V.I. Javorsky and L.K. Radugina.; Coal fires in the Kuznetsk basin and the related phenomena, *Gornyi zhurnal*, 1932, no. 10, pp. 55. [In Russian]

23. See: B.L. Isachenko and N.I. Mal'chevskaya. Biogennoe samorazogrevanie torfianoi kroshki, *Doklady* AN, 1936, T. IV, No. 8, p. 364.

24. See: A. Brehm. *Life of animals.* 4th edition, completely revised and considerably enlarged by Professor Otto Zur-Strassen. An authorized translation translation edited by a professor of the Psycho-neurological Institute for Women and the St. Petersburg M. N. Knipovich Medical Institute for Women. Vol 7. Birds. St. Petersburg, 1912. p. 15.

25. See: I.G. Frazer. op. cit.

This photo of the Aurora Borealis over Canada was taken by a member of Expedition 30 on the International Space Station

istence hundreds of thousands of years ago or more.

The surface of the planet was radically changed after that discovery. Everywhere sparkled, smoldered, and emerged a hearth of fire, wherever Man lived. On account of this discovery, he survived the cold glacial period.

Man created fire in the midst of living nature, subjecting it to combustion. In this way, by means of fires on the steppes and blazes in the forests, he received the power, relative to the vegetative and animal world surrounding him, which thrust him out of the ranks of other organisms, and presented itself as the prototype of his future existence. Only in our day, in the 19th and 20th Century did Man possess other sources of light and heat—electrical energy. The planet began to glow ever more, and we are presently at the beginning of a time, the significance and future of which for a time remains beyond our ken.

110.

There passed many tens, if not hundreds, of thousands of years until Man possessed other sources of energy, some of which, like steam-power, for instance, appeared to be the direct results of the discovery of fire.

In the course of long millenia, mankind radically transformed his role in the realm of living nature and in a fundamental way transformed living nature on the planet. This began already during the glacial period, when Man began to tame animals, but for many thousands of years this was not so clearly reflected in the biosphere. During the Paleolithic period, only the dog seemed to have a connection to Man.

A fundamental change began in the northern hemisphere beyond the boundaries of the glaciation after the retreat of the last glacier.

It was the discovery of agriculture, creating food independently of the bounty of untilled nature, and the discovery of breeding livestock, which, apart from its significance for Man's sustenance, accelerated the movement of Man.

Today it is difficult to determine precisely the conditions under which agriculture may have arisen. The natural environment surrounding Man at that time, some 20,000 or more years ago, was far different from that we see today in those same locations.[26] It is the result not only of a transformation of Man's cultural activity, as

26. It seems to me that the investigations of N. I. Vavilov regarding the centers where the domestication of animals and plants occurred will compel us to push back considerably further than 20,000 years ago the estimated date for the beginning of agriculture. N. I. Vavilov. *Problema proiskhozhdeniia kulturnykh ractenii*. Moscow-Leningrad, 1926.

was still not long ago believed, but also of a fundamental transformation of the environment of the glacial period in which we are now living. We can clearly see that even in the more recent historical period, the last 5-6,000 years, Man has experienced geological changes in the biosphere. The regions of China, Mesopotamia, Asia Minor, Egypt, possibly even regions of Western Europe beyond the limits of the taiga regions of those times, in terms of its climate, its aquatic regime and its geological morphology, were radically different from today, and it's not possible to explain this simply as a result of the product of the cultural work of Man and its inevitable, albeit unpredictable, consequences. Alongside the cultural labor of Man the spontaneous process of the freezing of the Glacial Maximum proceeded apace, increasing or decreasing in intensity, a process lasting some hundreds of thousands of years—the process of the anthropogenic age.

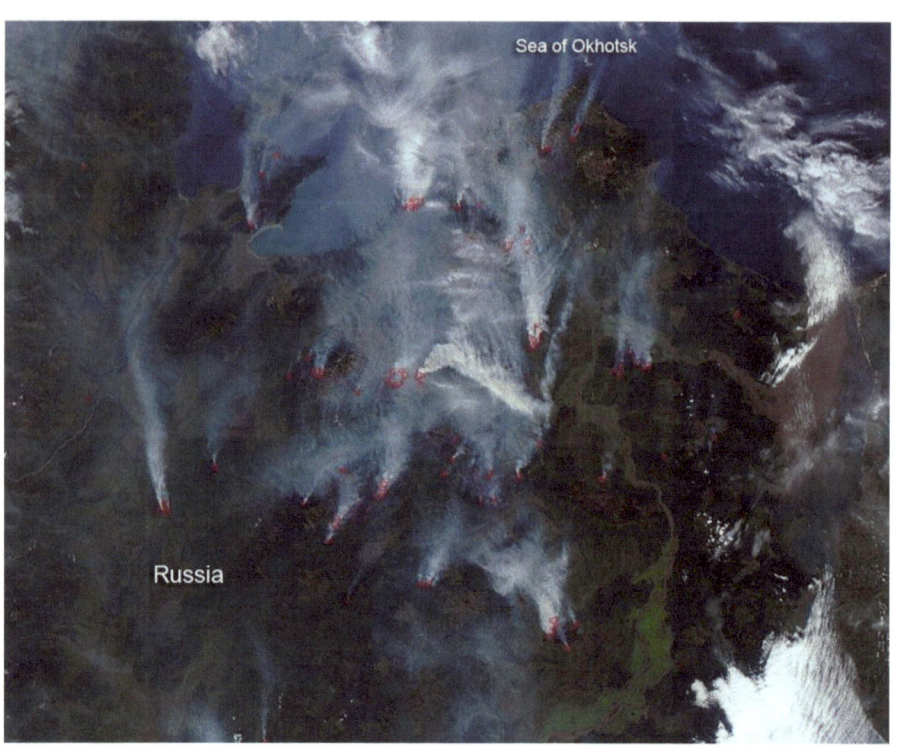

NASA

Bog land fire, burning since 2012 in Russia.

111.

With the present level of human culture, agriculture is not able to encompass the entire land surface of the Earth. In a recent [1929] estimate, the area of the land devoted to agriculture did not exceed 129.5×10^6 square kilometers, that is, 25.4% of the surface of the planet.[27] If we consider only the land area of the planet, this becomes 86.3%. We probably have to consider that figure exaggerated, but in general it gives an impression of the tremendous cultural biogeochemical energy by which mankind transformed, in the course of 20,000 years or more, the surface of the planet. We have to keep in mind the fact that the Arctic and Antarctic, the deserts and semi-deserts of north and southern Africa, Central Asia, and the Arabian Peninsula, the North American prairie, a significant portion of Australia, and the high plateau and the high mountains of Tibet and North America are either not suitable for farming or can be farmed only with great difficulty. Taken together, these make up not less than one-fifth of the land area of the globe. One must also note that for Man, even after the discovery of fire at the beginning of his cultural

labor, the taiga and the tropical forests represented nearly insurmountable barriers to agriculture. He would have to struggle a long time under these circumstances with the resistance mounted by insects and wild mammals, parasites and weeds, which devoured an enormous, and not infrequently, an overwhelming portion of the product of his labor. Even today in our agricultural endeavors, weeds envelop one-fifth to one-fourth of the harvest—in the beginning, that figure would certainly be a minimal one.[28] Nowadays we have, thanks to the socialist construction of our country, somewhat more accurate figures for calculating the intensity and the potentialities of this form of biogeochemical energy of mankind. We are undergoing an extraordinary expansion of cultivated land. As N. I. Vavilov and his colleagues have shown, only in the last two years (1930-31) the land under cultivation has increased by 18 million hectares, which would have required decades by the old standard.[29] With the aid of planned calculations being utilized by eminent specialists, a general map of our own country has been developed. It embraces an area equal to 2.14×10^7 square kilo-

27. H. Rew. Agricultural Statistics. *Encyclopaedia Britannica*, 1. London, 1929, p. 388.

28. A. I. Maltsev. "The most recent achievements in studying the weeds in the USSR." *Achievements and perspectives in the area of genetics and selection*. Leningrad, 1929, p. 381.

29. N. I. Vavilov, N.v Kovalev, N.S. Pereverzev. "Plant breeding in connections with the problems of agriculture in the USSR", *Rastenievodstvo*, vol. 1, ch. 1. Leningrad-Moscow, 1933, p. VI.

meters, or 16.6% of the Earth's land surface. Of this, 5.68x10⁶ square kilometers beyond the limits of its northern boundaries are unfit for cultivation. In total, there are around 11.85 x 10⁶ sq.km. unsuitable for agriculture, leaving 9.53 x 10⁶ square kilometers fit for cultivation. Thus the greater part of our country lies beyond the boundaries of modern agriculture or else is deemed unsuitable for it.[30] But this area may be significantly reduced through improvements. The government plan of ameliorating these conditions according to the calculations of L. I. Prasolov[31] will allow an increase of arable land by about 40%. Obviously this is not the end of the possibilities, and there is hardly any doubt that if mankind finds it necessary or desirable, he would be able to develop the energy needed to bring under cultivation the entire land surface, and perhaps even more.[32]

112.

We have still in China an intensive agriculture fully developed for generations, which, in a rather stationary state, existed for more than 4000 years in a country with a huge land area of more than 11 million square miles. Without a doubt, the country's topography changed during that time, but the system of production and agricultural customs were maintained and transformed the mode of life and nature. Only in the most recent period, in this century, does the mass of the population find itself in continual flux with customs that have lasted thousands of years being uprooted. We might speak of Chinese society as a purely agricultural civilization.[33] For countless generations, in the course of more than 4000 years, the population, in general remaining uninterruptedly in the same location, altered the country and in their social existence merged with the surrounding nature. Here probably the greater part of agricultural products are produced, and,

Vavilov Institute of General Genetics

N. I. Vavilov, plant biologist and geneticist, who was intent upon eliminating famine in Russia. He was targetted for elimination by Trofim Lysenko, and died in Saratov prison in 1943.

yet the population lives under the constant threat of famine.[34] More than three-fourths of the population are farmers. "A large part of China is an ancient nation, settled by farmers with the fields where they worked so close to their limits that large harvests were difficult to keep up. The roots of the Chinese go deep into the earth... The most significant element of the Chinese landscape is thus not the soil or vegetation or the climate, but the people. Everywhere there are human beings. In this old, old land, one can scarcely find a spot unmodified by Man and his activities. While life has been profoundly influenced by the environment, it is equally true that Man has reshaped and modified nature and given it a human stamp. The Chinese landscape is a biophysical unity, knit together as a tree and the soil from which it grows. So deeply is Man rooted in the earth that there is but one all-inclusive unity, not Man and nature as separate phenomena but a single organic whole."[35] And in spite of such unbroken, indefatigable labor of many thousands of years, only a little more than 20 percent of the land area of China is under cultivation,[36] while the remaining area in such a large and naturally rich nation might be improved through government measures, first made possible with our present level of scientific knowledge. A working population existing for thousands of years lives in an area of 3,789,330 square kilometers with an average capacity of 126.3 people on each square kilometer. That is almost the limit of the maximal use of agri-

30. L.I. Prasolov. "Land reserves for plant breeding in the USSR", Ibid, p. 31.

31. Ibid, p. 37.

32. The possibility of encompassing the oceans in one or another form was expressed a number of times in the scientific utopias even at a time when the physical power of Man was clearly negligible with regard to the powers of the oceans. In the interesting utopia of B. P. Weinberg [*Twenty thousand years from the beginning of the elimination of the oceans*, op. cit.] there is a discussion of that state of Man, which begins, when the reproduction of Man embrace the entire landmass—the state of the neutralization of the oceans. B. P. Weinberg assumes that in the 21st Century that question will be seriously broached. To a certain extent these questions have already become real issues before the mind of Man. The example of Holland in the past on a miniature scale and the idea of Goethe's Faust also on a small scale in the 18th and the beginning of the 19th Century already appear as realistic prototypes of the future. Now it is a matter of establishing a permanent, stationary floating base, existing outside any land area, in the midst of oceans and seas—this also merely the first inkling of what the future may bring.

33. F. Goodnow. *China, an Analysis.* Baltimore, 1926.

34. G.B. Cressey. *China's Geographic Foundations: a Survey of the Land and its People*, New York-London, 1934, p.101.

35. Ibid. p.1.

36. I am using figures provided by G. Cressey as to the total area of tilled land in the provinces and the fields of the small agricultural units and compared it with the total land mass of China. These figures are related to the period between 1928 and 1932. Rhe statistical review by Cressey (p. 395) for agricultural China (including the Hinggan Mountains, the Central Asian steppes and deserts and the areas adjacent to Tibet) gives the number as 379 square kilometers for a population greater than 477 million people—22% of the territory. Thus it is clear that the population is concentrated on a small land area which is utilized to its maximum.

cultural land. It will be, Cressey correctly indicates, from the standpoint of ecological botany, something of a final stage. "Here we have an old stabilized civilization which utilizes the resources of nature to the limit. Until new external forces stimulate change, there is but little internal readjustment." [37]

"The Chinese landscape is vast in time as well as in area, and the present is the product of long ages. More human beings have probably lived on the plains of China than on any similar area on Earth. Literally trillions[38] of men and women have made their contribution to the contour of hill and valley and to the pattern of the fields. The very dust is alive with their heritage." [39] That four thousand-year culture, before it adopted this stabilized form must have experienced a more grim and tragic past, since the former conditions of nature in China were completely different, enveloped by a totally different milieu, with humid forests and marshes; to subdue these and bring them under cultivation—destroying the forests and ridding them of their animal inhabitants—would have taken tens of thousands of years. The latest discoveries reveal that while Europe was experiencing the movements of glacial ice, in China there developed a culture under conditions of a pluvial period.[40] Certainly, the basic system of irrigation, to which agriculture in China owed its existence, had its roots far back in history, 20,000 years ago and more. Until the end of the 20th Century such an ecosystem may have existed in a certain equilibrium. But it could exist only because China was, to a certain degree, isolated, and from time to time the population was decimated by wars, hunger, famine and floods; irrigation work was too weak to cope with such mighty rivers as the Yellow River. Now all of that is rapidly becoming a thing of the past.

In China we see the last example of a unifying civilization lasting millenia. We see at the beginning of the 18th Century when Chinese science stood in high esteem, it experienced a historic shift, and missed the chance of being incorporated at the necessary moment into world science. It was included there only in the second half of the 19th Century.

113.

Agriculture would appear as a geological force, transforming the surrounding nature, only when it occurred together with the raising of livestock, namely, when Man, in addition to the selection and cultivation of plants necessary for his sustenance, also selected and began to breed the animals he needed. Man accomplishes this geological work inadvertently, stimulating a greater reproduction of a certain species of plant and animal organisms, always creating for himself an available supply of food and maintaining a food supply for the animals he needed. In raising livestock he not only obtained a guaranteed food supply, but also increased his muscle-power, allowing him to put more fields under agricultural production at an earlier stage.

In the work of the livestock, he obtained for himself a new form of energy, enabling him to support a larger population, create large settlements, an urban culture, and free himself from the otherwise ever-present threat of famine.

In doing this, he always remained within the bounds of living nature.

During the past centuries, in our age of steam and electricity, the labor-power of livestock and the muscular energy of animals and Man, begin to play a secondary role in the expansion of agriculture. However, even with that, Man does not transcend the bounds of living nature, since the primary source of the electrical and steam energy is that same living nature in the form of living matter or, even more now, past living organisms which have been transformed through geological processes. This energy is obtained from coal and oil. After all, Man has in this manner always made use of solar radiation, which illuminated the Earth for hundreds of millions of years before his appearance, and, which, transmitted through living organisms, were utilized by him either directly, or as preserved in their fossilized form.

In agriculture and livestock are manifest more than anything the cultural biogeochemical energy directed by reason, creating for Man new conditions for his habitat in the biosphere. By these means living nature by and large was radically transformed. For many tens of thousands of years, the inert matter of the biosphere was affected by Man to a degree barely comparable to the present profound transformation in his surrounding living habitat.

As a result of this there has been created a new face of the Earth, that in which we are now living and which began to emerge only in the last millenium. Now change occurs ever more abruptly with each passing decade.

But agriculture alone, even without livestock, radically transforms nature. For in the living nature surrounding it, every vacant area is filled by living organisms, and in or-

37. Cressey, op.cit., pp.1-2

38. Certainly it is not a question of trillions but of a considerably large number of people that have lived on the territory of China, as the presence of the human species and its predecessor, Sinanthropus, in that territory, would have been established in the course of hundreds of thousands of years. The appearance of a new species or race, powerful enough to provide the beginnings of modern Man, may have occurred in one family or at one stage, but could also have occurred over a rather large area. But even in the first case, the number of offspring originating from a single couple must have been much greater than 10 billion over the course of hundreds of thousands of years (even if you introduce) corrections to the general forebears of particular individuals. On this, see: E. Le Roy, Op.cit.

39. Cressey. op.cit., p. 3.

40. As for ancient China see: M.Granet. *La civilisation Chinoise*. Paris, 1926, p. 82 ff.

Chinese agriculture remained static for many centuries. Here, a representation of a farmer tilling his field with the help of his cattle.

der to introduce new life, Man must make a place, clearing the land from other living organisms. Moreover, he must continually maintain that new living substance established by him from the surrounding pressure of other life, from the animals and plants, which were inserting themselves into the vacant areas cleared by him. He has also to preserve the fruits of his labor from animals and plants, lest they be devoured by mammals, birds, insects, fungi, etc. We see that even at the present stage, he has not been able to definitively accomplish this.

Agriculture together with livestock, continually maintained by human thought and labor, in the end fulfills an enormous geological task. Old life is destroyed, and new life is created—new species of animals and plants, created by the thought and labor of Man, emerge from the old, created under different conditions. But even the world of wild animals and plants that have not been directly affected by Man are inexorably transformed in

the new conditions of life created by the biogeochemical energy of Man.

114.

The raising of livestock, apart from agriculture, produces a tremendous change in the surrounding living nature. For it consumes food and condemns to a slow or rapid extinction large mammals, among which Man selects a few species. Man appeared at the end of the Tertiary Period, in the epoch in which large mammals reigned in the biosphere, as Osborne correctly points out. [41]

At the present time, it can be said that these mammals have either practically died out or are rapidly disappearing, and are preserved only in reserves and parks, where their number is relatively stable. Observations in these large reserves show that here practically always a stable dynamic equilibrium is achieved, even independently of Man's will, in which reproduction is regulated by the limited quantity of food for the herbiverous animals and by the quantity of carnivorous animals, for which these serve as food.[42] [43] With an insufficiency of food—and a weakening of their organisms, reproduction is primarily determined by diseases caused by microorganisms. But the total preserved numbers of wild herbiverous mammals can certainly not be compared to the number of domesticated animals—horses, sheep, cattle, pigs, goats, etc., and it is conceivable that their number in the Tertiary Period hardly surpassed the number of domesticated mammals. We don't know that number with any exactitude, but we do have some idea about it. At the present time it exceeds by hundreds of times the number of the human population. According to M. Smith (1910), at the beginning of the century, the number equalled 138 billion. According to H. Rew[44] the number in 1929 for horses, bulls, sheep, cows, and pigs reached 15.7 billion. Not taking into account here species of domesticated animals does not change the order of magnitude of the number. Thus one might say that the expression in billions fluctuated between 16 and 138 billion, significantly exceeding the number of people. The number fluctuates sharply as it is under human control. Thus, according to J. Dufrenoy[45], from 1900 to 1930, the number of livestock diminished

41. H. Osborn. *The Age of Mammals in Europe, Asia and North America.* N.Y., 1910.

42. See: James Stevenson-Hamilton. *South African Eden: from Sabi Game Reserve to Kruger National Park.* London, 1937.

43. M. Smith. Agricultural Graphics. United States and World Crops and Live Stock, *Bulletin of United States Department of Agriculture*, Washington, 1910. N 10, p. 67.

44. H. Rew. *Encyclopedia Britannica*, v. 1, 1929, p. 388.

45. J. Dufrenoy, *Revue générale des sciences pures et appliquées.* Paris, 1935, N 46, p. 72.

by one fourth, displaced by machinery. As Man came to possess new sources of energy, that number quickly dwindled before our eyes, as, for example, the number of horses, donkeys and mules, owing to the increase of tractors and automobiles.

115.

The appearance of livestock and agriculture was established at various times and in various locations within the span of 20,000 to 7,000 years ago, gradually increasing in intensity as we approach our own era. The transition from the nomadic (migratory) hunting and food-gathering phase to our pres-

Cuneiform writing of the ancient Sumerians.

ent settled mode of life based primarily on agriculture, occurred at various periods on the boundaries of the uninhabited zones of the temperate latitudes stretching from present-day Morocco to Mongolia. This was possibly the result of climatic changes after the retreat of the last glacial cover and the weakening of the pluvial period[46] of the Pleistocene.

Seven or eight thousand years ago appear the first powerful states based on agricultural and the first large cities. This provided Man with the possibility of unimpeded reproduction with only minor interruptions. Here were established the urban civilizations of the Celtic and the Berber states and their predecessors in Egypt, Crete, Asia Minor, Mesopotamia, northern India, and China. We are entering the age, which power and significance is steadily and rapidly growing over the last three centuries (and from which legends have been preserved and have come down to us, and countless material relics provided, unearthed through archeological excavations).

You might say that within the last five to seven thousand years the continuous creation of the noösphere has proceeded apace, ever increasing in tempo, and that the increase of the cultural biogeochemical energy of mankind is advancing steadily without fundamental regression, albeit with interruptions continually diminishing in duration. There is a growing understanding that this increase has no insurmountable limits, that it is an elemental geological process.

116.

It is appropriate to add here a few additional facts. It is possible to date to somewhat earlier than 4,236 B.C.[47] or earlier the origins of the Egyptian calendar (which is based on many years of observations of Sirius), which served as the basis for the chronology of the entire ancient world right up to the present moment, where it is found spread throughout the entire noösphere. Even before that, somewhere between 5-4,000 B.C., there existed an urban culture in India, Mesopotamia, and Asia Minor, with such a level of technology, which we even as recently as a few years ago did not suspect, encompassing a population numbering perhaps in the millions. At the end of that period, around 3,000 B.C., began a shift toward using animals for transportation, and in the course of the next 1,500 years this rapidly expanded, and included oxen, camels and horses. Around 3,300 B.C., in the temples of Mesopotamia, written script was being used. Records were kept by means of a complicated pictographic script, and around 1,600-1,500 B.C., the Semites in the Near East discovered the use of the alphabet. We can assume that around 2,500 B.C., we already had a clear manifestation of scientific thought, and around 2,000 B.C. in Mesopotamia, we had the discovery of the decimal system. At this time old records, written some centuries earlier, were copied and preserved in libraries. Between the 15th and 16th centuries B.C., we note wide-ranging exchanges in

46. N. Nelson recently provided a concise review of this on a global scale. See: N. Nelson. Prehistoric Archeology; Past, Present and Future, *Science*, 1937, v. 85, N 2195, p. 87.

47. It may be that the choice is between these two dates—4236 B.C. and 2776 B.C. From what we now know, taking into consideration the growth of studies in history and archaeology, it appears that the first figure is the correct one. See: Naum Idelson, *The History of the Calendar*, Leningrad, 1925.

Russian Academy of Sciences

Vernadsky with the workers of the Biogeochemical Laboratory which he established in 1928

the cultural world of scholars, philosophers, and physicians of that period. Around 2,000 B.C. or earlier we have the discovery of bronze, probably simultaneously in several places, and around 1400 B.C., the discovery of iron, which in the course of a few centuries came into general use.

With these momentous achievements we have now arrived at the first century B.C., in which scientific, philosophical, artistic and religious creativity achieved an enormous development and laid the first foundations for our civilization.

In the course of the last 500 years, from the 15th to the 20th Century, Man's powerful influence over his surrounding nature and his comprehension of it, ceaselessly advanced, becoming ever stronger. In this period the entire surface of the planet was embraced by a single culture: the discovery of printing, knowledge of all earlier inaccessible areas of the globe, the mastery of new forms of energy—steam, electricity, radioactivity, the mastery of all the chemical elements and their utilization for the needs of Man, the creation of the telegraph and the radio, the penetration into the Earth to the depth of a kilometer by boring, and the ascension of men in aerial machines to a height of more than 20 kilometers from the surface of the Earth, and of mechanical devices, to a height of more than 40 kilometers. Profound social changes, giving support to the broad masses, advanced their interests into the first rank, and the question of eliminating malnutrition and famine, became a realistic option that can no longer be ignored.

The question of a planned unified activity for the mastery of nature and a just distribution of wealth associated with a consciousness of the unity and equality of all peo-

ples, the unity of the noösphere, became the order of the day. It is not possible to reverse this process, but it bears the character of a ruthless struggle, which, however, is grounded on the deep roots of an elemental geological process, which may last two or three generations, or more (although it is hardly probable judging from the tempo of evolution in the last thousand years). In that transitional stage, amidst the intense struggles which we are now undergoing, it would as well seem less likely that there will be any protracted interruptions in the ongoing process of the transition from the biosphere to the noösphere.

The scientific grasp of the biosphere which we now observe is an expression of that transition.

Its non-fortuitous nature and its connection to the structure of the planet—its outer envelope[48]—we must later subject to a possibly deeper thoughtful logical analysis, in considering an understanding of biogeochemistry.

All the above exposition is the result of precise scientific observation, and insofar as this was faithfully done, it ought to be considered a scientific generalization.

It is a scientific description of a natural phenomenon, without the assumption of any hypotheses, theories or extrapolations.

117.

Observing thus the developed scientific disciplines, we see that there exist sciences of different types: in the first category, we have those whose objects, and consequently whose laws, encompass all of reality, such as our planet and its biosphere, as well as the cosmic expanses—that is, sciences whose objects correspond to the fundamental, universal phenomena of reality. The second category is related to phenomena which are characteristic of our Earth.

---

48. Actually this is possibly a second envelope of the Earth's crust—the stratosphere, encompassing life (mainly through Man—the noösphere), and it ought to be included in the biosphere. (See:V.I. Vernadsky, "On the limits of the biosphere," *Izvestiia* AN, geological series, 1936, No. 1, p. 3-24). We should think of the upper layers (60-1000 km.) not as part of the Earth's crust, but as analogous to the Earth's crust in the division of the planet, that is, concentric with the planet. The Earth's crust will be the second sphere, and the biosphere is its outer envelope. This, of course, will soon become clear.

In this latter category, we might theoretically admit two classes of scientific objects to be investigated: general planetary phenomena, and individual, purely terrestrial, phenomena.

At present, however, it is not always possible to differentiate reliably and with a sufficient degree of certainty between these two cases. This remains a task for the future.

Here it concerns all the sciences of the biosphere, with the sciences of the humanities, with the Earth sciences—botany, zoology, geology, mineralogy—in all their scope.

Considering such a condition of our knowledge, we can distinguish an expression of the influence on the structure of the noösphere of two areas of human thought: the sciences common to all reality (physics, astronomy, chemistry, mathematics), and sciences related to the Earth (biological, geological, and humanistic sciences).

118.

Logic occupies a special position, in the most intimate manner connected to human thought, embracing equally all of the sciences: both the humanities, on the one hand, and the mathematical sciences, on the other.

Actually it should be included in the realm of planetary phenomena, because only by means of it is Man able to understand and scientifically grasp all reality—the scientifically structured Cosmos.

Scientific thought is both individual and social. It is inseparable from Man. Not even in his deepest levels of abstraction can an individual transcend the realm of his existence. Science has a real existence, and like Man himself, is most closely and inextricably bound to the noösphere. The individual is obliterated—"decomposed"—when he goes beyond the logical grasp of his intellect.

But the mechanism of the understanding, tightly linked to speech and concepts—the logical structure of which is complex, as we shall see (observe the digression on logic at the end of the book)—does not encompass the totality of Man's knowledge of reality.

We see and we know, but we know in an everyday, not in a scientific way, that creative scientific thought transcends the bounds of logic (including logic and dialectics in its various forms). The individual, in his scientific accomplishments, bases himself on phenomena, which are not encompassed by logic (however broadly we understand that term). Intuition, inspiration, the basis of the greatest scientific discoveries, proceeding and operating further in a strictly logical manner—is not brought forth by either scientific or logical thought, nor is it connected to words or concepts in its genesis.

With regard to this fundamental area in the history of scientific thought, we are entering into a realm still not fully grasped by science. But not only can we not take it into consideration, rather we must increase our scientific focus on it. At present this area of philosophical speculation is somewhat clarified, but in general still finds itself in a chaotic state.

This area has been investigated with greater interest and depth in Hindu philosophy, both ancient and modern. Here we have attempts to delve into this realm, barely touched by science.[49] How far it will conduct human thought, and give it a direction—of this we have no definite knowledge.

We only see that a large realm of phenomena, which possess a rigorously lawful, most intrinsic, relationship to the social order, and ultimately, to the biosphere—even more to the noösphere—namely, the world of artistic creation, is not reducible in any meaningful way in any of its parts, for example, in music or architecture, to verbal representation, and yet it exerts a great influence on the scientific analysis of reality. The mastery of this cognitive apparatus, little reflected by logic, is a task for the future.

119.

Biogeochemistry in its greater part, the objects of which are atoms and their chemical properties, ought to be ascribed to the category of the general sciences. However, as a sub-division of geochemistry, that is, the geochemistry of the biosphere, it appears as a science of the second type, that is, associated with the small, more circumscribed, natural bodies of the universe—with the Earth, or, in the more general case, with the planet.

But, in studying the manifestation of atoms and their chemical reactions on our planet, biogeochemistry fundamentally transcends the limits of the planet, and basing itself, like chemistry and geochemistry, on the atom, it is thereby involved with more potent problems than those simply characteristic of planet Earth—namely, with the science of the atom and with atomic physics—with the very foundations of our understanding of reality in its cosmic dimension.

This is less clear with regard to life which is studied by it in its atomic aspect.

Do the problems of biogeochemistry also here transcend the boundaries of the planet? And how far out do they emanate?

---

49. To avoid any misunderstandings I should explain that I have here in mind not some theosophical quest, which would be far removed from contemporary science, as well as from contemporary philosophy. Both in the new and in the ancient Indian thought there exist philosophical currents, by no means contradicting our contemporary science (actually less so than do the philosophical systems in the West), as, for example, some systems associated with Advaita-Vedanta, or even the religious-philosophical investigations, as far as I know them, of the prominent religious thinker, Aurobindo Ghoshi.

The "Greening" of Vladimir Vernadsky:

How The Russellites Sabotage Science

by William Jones

While the name Vladimir Vernadsky is still not as widely known here in the United States as it should be, given his prominence as one of the greatest scientific thinkers of the last century, the prevalent view of Vernadsky is largely based on a fraud perpetrated by the acolytes of that Malthusian genocidalist, Bertrand Russell, whom economist and statesman Lyndon LaRouche so aptly labeled the most "evil man in this century." To the extent Vernadsky is known within the American scientific community, he is largely seen as some sort of early ecological guru. The fraud of this view, tragically, has also become prevalent within Russia itself, where there is less excuse for it, as Vernadsky's works have been widely publicized in his native language. His name is often equated with that of wacko Gaia worshipper, James Lovelock, who belatedly also labeled himself a "Vernadskyian," although Vernadsky's world-view was, in fact, diametrically opposed to that Greenie mystic.

While Vernadsky was a natural scientist, who provided a solid scientific basis to the notion of the "biosphere," so much abused these days by the lunatic Greens, he saw the productive activity of man, a result of the biosphere, but transforming it into a higher state, as the most important element in its continued development. The stage of the biosphere characterized by the intellectual activity of man Vernadsky called the noösphere (*noös* is Greek for *mind*). Unlike the Greenies who believe that mankind should shut down its industrial activity in order to become "one with nature," Vernadsky believed that it was precisely man's creative ability to develop his technology, to develop new ideas resulting in productive breakthroughs, that provided man with essentially "unlimited resources." While insisting that such advances be implemented with scientific rigor, he was invariably opposed to placing restrictions on continued technological progress. Indeed, without such progress, Vernadsky knew the human race would quickly be on the road to extinction.

Now on the occasion of the 150th anniversary of his birth, it is fitting that we set the record straight and expose the fraud which has been imposed on an unknowing public by the Greenie acolytes of Russell and his cohorts.

Who Was Vladimir Vernadsky?

Vladimir Ivanovich Vernadsky was born in 1863, the son of economist Ivan Vasilievich and Anna Petrovna Vernadsky. The elder Vernadsky had been instrumental in the movement which led to the freeing of the serfs by Alexander II in 1861. He was also instrumental in introducing the works of the anti-Malthusian American economist Henry Charles Carey to the Russian intelligentsia, works which caused great enthusiasm among leading Russian economic circles. Carey's writings were rapidly

Archive Collection, Russian Academy of Sciences

Over the door of his office, Vernadsky kept the picture of George Washington that had always hung in his boyhood home.

translated into Russian. Young Vladimir, however, was more attracted to science than to economics. A portrait of George Washington graced his boyhood home, and later, the same portrait hung in his laboratory office. Abraham Lincoln was characterized by Vernadsky as a "hero for all times," paraphrasing a famous work by Russian writer, Mikhail Lermontov, "A Hero for Our Time." Vernadsky became acquainted at an early age, thanks to his father, with the work of the great 15th century scientist, Cardinal Nicholas of Cusa, whom Vernadsky, as a young professor, would laud in his lectures on the history of science, as the founder of modern science, leading into the Renaissance:

> One of the predecessors of the ideas of Copernicus was Cardinal Nicolas of Cusa (1401–1464) to whom I have previously referred. The son of German peasants, a faithful and passionate representative of the Catholic Church, he was one of the most original and prodigious minds of his time. In his works we find the seeds of a variety of ideas that have since become a part of contemporary thought. He died in 1464 soon after the discovery of printing, and his works were left in manuscript form, threatened with the same fate as was common with many of his predecessors, becoming known only much later, when all direct living contact with them had disappeared. But the works of Cusa avoided this fate. He was published 40 years after his death, but before his direct influence had waned. The first (now extremely rare) edition appeared in Rome in 1501. It was the first appearance in human thought since the ancient Greeks of the representation of the Earth turning on its axis, and revolving around some point in space, which Cusa considered to be, not the Sun, but rather a certain pole of the Universe... We see

Archive Collection, Russian Academy of Sciences

Vernadsky, here in Prague in 1926, cannot cease to examine that phenomenon of life that so engaged his life's work.

Archive Collection, Russian Academy of Sciences

Vernadsky (on the right) photographed here together with other members of the left faction of the Russian Duma.

everywhere the influence of these ideas of Cusa, with which Copernicus was also acquainted. The significance of the works of Cusa was also seen in other areas of thought as well, and his the works are continually cited, primarily by the more innovative spirits, throughout the course of the 16th and 17th centuries.[1]

Studying the work of Alexander von Humboldt, particularly Humboldt's epic summary of the science of his day, *Cosmos*, Vernadsky devoted himself to the field of science as his best means of contributing to the progress of man, specializing in mineralogy and soil science, and later geochemistry. While maintaining a clear political engagement all his life, he felt that the progress he was making in the development of science represented his greatest contribution to his country and to the world. When the Bolsheviks took power in Russia, Vernadsky, one of the founders of the Constitutional Democratic (Kadet) Party traveled to Ukraine, still under the Whites, in order to avoid arrest. When he finally decided in 1921 to return to work in Bolshevik Russia where the Kadet Party was now banned, this put an end to any direct political activity on his part, although he would exert a great deal of influence with regard to science policy in the Soviet Union. Making his major discovery in the early 1920s of the inexorable role of life in the development of the Earth's surface, Vernadsky went on to make major breakthroughs in a variety of related fields, particularly in mineralogy and soil science, and created an entirely new field of science—biogeochemistry. Vernadsky also became the first person in Russia in the 1920s to lobby for a major research program for developing atomic energy.

1. Vernadskii, V.I. "Izbrannye trudy po istorii nauki" *Nauka*, Moscow, 1981. p.101.

Vernadsky is credited with the most comprehensive elaboration of the notion of the biosphere. His discovery of the unique quality of life to rapidly envelop over an entire area of the globe once it appeared on the scene, came to Vernadsky in his self-imposed exile in his beloved Ukraine during the period of the Russian civil war in the early 1920s. Vernadsky was astonished at first by the speed with which life proliferated and he took it upon himself to measure that rate. In the chaos of the Russian political world following the Bolshevik Revolution, Vernadsky also found solace and hope in his discovery of this elemental force of life to rapidly expand and proliferate, a force which he felt ultimately characterized the universe as a whole, including man's consciously directed social and economic development. Later, in 1939, Vernadsky would write:

It is evident that the phenomenon of the expansion over the entire surface of the planet by a single species developed broadly in the case of aquatic life such as microscopic plankton in lakes and rivers, and with some forms of microbes, essentially also aquatic, on the thin film of the Earth's upper surface, and was disseminated through the troposphere. For larger organisms, we observe this almost in full measure with certain plants. For man this begins to appear in our time. By the 20th Century the entire globe and all the seas have been encompassed by man. With the rapid progress of communications, mankind is able to maintain continual contact with the entire world, and in no place is he alone or helplessly lost in the immensity of Earth's nature.[2]

In the same way that life becomes a predominant force in the lithosphere, bringing to it new processes which

Much of Vernadsky's legacy lies in numerous manuscripts now preserved by the Russian Academy of Science.

enrich and enhance it, so too does man's productive activity become an element in the biosphere, enriching and enhancing its productivity. This was characterized by the increase of energy throughput occasioned by man's activity and by the ability of man to support ever more efficiently an ever-increasing population. This is due to man's development of technology, a result of his noetic activity. And, placed on the cusp of a new century with the discovery of the atom, Vernadsky felt that the rate of development of technological progress was exponentially increasing. Writing in the 1930s, Vernadsky states:

In the course of the last half millenium, from the 15th to the 20th Century, the development of man's strong influence over his surrounding nature and his comprehension of it, continued apace, growing ever more powerful. During this period the entire surface of the planet was encompassed by a single culture: the discovery of printing, a knowledge of all earlier inaccessible areas of the globe, the mastery of new forms of energy—steam, electricity, radioactivity, the mastery of all the chemical elements and their utilization for the needs of Man, the creation of the telegraph and the radio, the penetration into the surface of the Earth to the depth of one kilometer by boring, and the ascension of men in aerial machines to a height of more than 20 kilometers from the surface of the Earth, and of mechanical devices, to a height of more than 40 kilometers. Profound social changes, having been given support by the broad masses, thrust their interests into the first rank, and the question of eliminating malnutrition and famine, became a realistic option that could no longer be ignored.[3]

These words of Vernadsky are a far cry from any "Green" manifesto, which one would expect from his depiction as a proto-ecologist.

Vernadsky's Outlook

Vernadsky was well aware that his new conception of the biosphere was a ground-breaking one. He also knew that it required a larger audience in order to achieve its full import. While his first major work on the topic, *The Biosphere*, was quickly translated and published in French in 1929, the publication of his other writings would take a longer time to appear in translation, if at all, particularly with regard to an English translation. By the 1930s, Vernadsky was working on a series of papers, under the general title "Problems of Biogeochemistry," which summarized his mature views on the role and meaning of the biosphere and on man's increasingly preponderant role

2. Vernadsky, V.I. "Scientific Thought As A Planetary Phenomenon," *21st Century Science & Technology*, Spring-Summer 2012. p. 19.

3. *Ibid.*, p. 30.

G. Evelyn Hutchinson was a member of that stable of characters who followed Bertrand Russell's population-control agenda.

in its development. He was particularly anxious to have these papers published in English, to make the English-speaking scientific world fully aware of his new conception of man and the universe.

The presence in the United States of Vernadsky's son, George, and of his daughter, Nina, both of whom had emigrated after the Bolshevik Revolution, put them in an ideal position to foster knowledge of their father's work here, even as it was taking shape in Russia. Also at Yale was a young Russian professor, Alexander Ivanovich Petrunkevitch, the son of one of Vernadsky's political mentors and close collaborators in the Kadet movement, Ivan Ilyich Petrunkevitch. Alexander Ivanovich had also been a former student of Vernadsky, and, after his emigration, became a zoologist at Yale, specializing in the study of spiders.

George Vernadsky was a professor of history at Yale University. Also at Yale was a British geologist and limnologist named G. Evelyn Hutchinson. Hutchinson was something of a typical by-product of the inter-war period at Britain's institutes of higher learning, particularly at Cambridge, where Hutchinson received his education. This was at the time a hotbed of Darwinism, Malthusianism, and philosophical reductionism. Names like Julian Huxley, J.B. Haldane, Bertrand Russell, anthropologist Gregory Bateson, as well as novelist, H.G. Wells, are prominent in this context. Bateson and Haldane were particularly close friends of Hutchinson at Cambridge. What united this crowd was their commitment to Darwinism and to a neo-Malthusian world outlook, which has always remained at the heart of the British imperial world-view.

Vernadsky early realized that Malthus's predictions were fundamentally flawed.

The position of Malthus, the classic spokesman of zero population growth, is too well known to dwell on here. But also Charles Darwin, who essentially viewed man as another form of beast, somewhat like a clever ape, took his cue from the work of Malthus. As he himself admits, it was a reading of Malthus's *An Essay on the Principle of Population* which prompted Darwin to compose his *Origin of Species*. Vernadsky had during his student days encountered the work of Pastor Malthus on population, and rejected it outright. Referring to Malthus' fundamental thesis, Vernadsky writes:

> Malthus doesn't realize that his fundamental results lead to entirely different conclusions. You might say that they are simply not true, because he did not take into consideration the fact that, estimating accurately the long-term growth of human population geologically, as regards food and the necessities of life, the expansion of plant and animals comprising it, must inevitably increase with greater force and speed, expressing a *more rapid* rate of reproduction, than that of the population. It's necessary to always have this correction in mind. Historically, it is only the irrational elements in our social system that make it difficult to clearly observe the effect of this natural phenomenon.[4]

Man is capable of creative thought, said Vernadsky. And thanks to this capability, he succeeded in developing in the material world around him new sources of energy, the latest example of which, in Vernadsky's day, was atomic energy. Because of this unique noetic capability, man succeeds in moving to energy sources ever more potent, ever more dense, from fire, to coal, to oil, to nuclear. The development of man is characterized, therefore, by increasing energy-density, or more specifically, energy-flux density. Because of this creative ability, man, in contrast to all other species, was not facing limits to growth, but was capable of continually

Hutchinson created the field of "population ecology" which treated man as simply another animal species.

4. Vernadskii, V.I., *Khimicheskoe stroenie biosfery zemli i ee okruzheniia. Nauka.* Moscow. 2001. p. 302. (emphasis added)

developing new resources which could support an ever-expanding population. Verandsky's rejection and unequivocal refutation of the arguments of Pastor Malthus early in his career was no aberration, as the Greenies would have it, but rather the hallmark of his fundamental philosophical and scientific outlook.

The Russellites

But Malthusianism was something of an endemic philosophy for the British Empire, dedicated to the preservation of its hegemony over world political and economic developments, and was widespread at places like Cambridge and Oxford. One of the key representatives of the Malthusian viewpoint was Bertrand Russell, who touted himself a mathematician and philosopher. Never one to conceal his views, Russell was quite open about his genocidal policies. Writing in a 1954 article published in *Crux*, the journal of the Union of Catholic Students of Great Britain, entitled "Birth Control and World Problems," Russell explains his view:

Bertrand Russell's genocidal policies made him in the words of Lyndon LaRouche "the most evil man of the century."

Opponents of birth control make much of possible improvements in agricultural production either by new methods or by irrigation of deserts. What they refuse to face is that there is a limit to what can be done in this way, whereas there is no limit to the increase of geometrical progression. If the population of the world were to continue to increase at a constant rate, however slow, there would in time be only standing room, and no land whatever would be left for food production. Sooner or later therefore the increase of population must cease. Shall the cessation be brought about by war, by pestilence, or by starvation? No other possibility exists for the opponents of birth control—unless indeed, they were to advocate large-scale sterilization, which they find even more abhorrent.

Later, Russell (an early proponent of nuclear war against the Soviet Union) and his circles would help to spread the virus of his misanthropic world-view to an entire generation of Soviet scientists under the aegis of such "collaborative" "scientific" organizations such as the Pugwash Conferences and the International Institute for Applied Systems Analysis (IIASA) in Laxenburg, Austria. Russell would utilize the danger of "nuclear winter" in order to brainwash scientists about the need for a no-growth, "green" agenda. While Vernadsky was not alive when Russell wrote that particular tract, he was

quite aware of the general nature of Russell, who during the 30s was touting himself as an interpreter of the "philosophical implications" of Einstein's relativity theory. Writing in his diary in 1938 with regard to A.E. Fersman, a protégé and collaborator, whom he often chided for his lack of political courage, Vernadsky commented: "A.E. belongs to that type of scientist who feels *his* view of nature is so great, that he does not notice the paltriness of that 'view' when juxtaposed to the real greatness of nature itself, like B. Russell."

But Russell's views were rather mainstream for British intellectuals of an "imperial" outlook. And G. Evelyn Hutchinson was a man of the same mold. So at Yale, something happened to the project of publishing Vernadsky's works in English. Hutchinson was given a major role in the editing of Vernadsky's writings. Hutchinson created a field of dubious scientific worth called "population ecology" or "mathematical ecology." While his scientific work in that field was largely directed toward the populations of animal species, he, like Darwin, extrapolated his findings in the animal world to the world of man, warning that limits must be imposed on the growth of the human population. His "niche theory" of evolution described how each species, including man had to find its "niches" in this world of competition for Lebensraum and resources. Sadly, however, each species was relegated to its own particular "niche," beyond which it could no longer progress. As a professor at Yale, Hutchinson would go on to create a whole gaggle of ecology freaks, including biologist E.O. Wilson, Thomas Lovejoy of the World Wildlife Foundation, and many, many others. Because of his widespread influence, Hutchinson is characterized as the "father of ecology," although he himself attributed that title rather to Charles Darwin.

The Fraud

Already during his time at Yale in the 1930s, Hutchinson had learned of the work of Vernadsky, probably from his friend and colleague Alexander Petrunkevich. While Hutchinson didn't know any Russian, he had obtained a copy of the 1929 French edition of Vernadsky's *The Biosphere* and had his students read sections of it in his class. Hutchinson saw the possibility of using aspects of Vernadsky's work for his own purposes while suppressing Vernadsky's own world-view. Given Hutchinson's reductionist view of man, Vernadsky's idea of the noösphere and the role of human creativity in overcoming "limits to growth," even reflected in the more focused monograph, *The Biosphere*, was absolutely anathema to him.

Hutchinson had studied limnology at Cambridge during the post-World War I period in England, where the eugenicist movement was having its heyday. Evelyn imbibed his zero-growth philosophy literally from mother's milk. His mother, Evaline, a radical feminist, was an early adherent of sex psychologist Havelock Ellis, and a close friend of eugenics matron, Margaret Sanger, who fled to England from the United States to find more fertile ground for her anti-human philosophy.

The Hutchinsons were an integral part of the Cambridge social circle, which included the Darwins, the Huxleys, the Batesons and the Haldanes. At Cambridge, Hutchinson would strike up a close relationship with J.B. Haldane, who would later provide the backing of Western science for the checkered career of Alexander Oparin, the chief antagonist of Vladimir Vernadsky's views in post-war Russia.[5] Here he also struck up a friendship with Gregory Bateson, with whom he would collaborate at Yale in laying the basis for the counter-culture movement of the 1960s. When Bateson hooked up with the American social anthropologist Margaret Mead, Hutchinson would also become her friend and mentor, and in fact, her copy editor. Hutchinson was also close to British author and radical feminist, Rebecca West, who was for a time the wife of H.G. Wells.

Hutchinson received a professorship at Yale in 1928 and Yale would ever remain the lair from which he would spin his web of deviltry and deceit. He also served, together with Mead, on the staff of the American Museum of Natural History in New York. Hutchinson, Mead, and Bateson, as well as cultural anthropologist Ruth Benedict, would all participate in the conferences organized by the Josiah Macy Foundation in 1946, which were instrumental in creating the basis for the "alternative lifestyles" that would be foisted on America in the latter part of the 1960s, in the aftermath of the assassination of President Kennedy.

Editing Vernadsky

It was undoubtedly his connection with Petrunkevitch that brought Hutchinson into a position to influence the Vernadsky "legacy" in the U.S. Hutchinson, now retooling himself from limnology, the study of lakes, to bio-

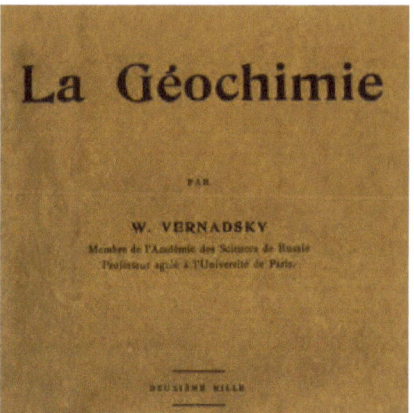

Vernadsky's lectures on geochemistry at the Sorbonne were published in French in 1924.

chemistry, was, because of his "expertise" in the field, given the task of editing George Vernadsky's translation of two of his father's papers in a series Vernadsky labeled "Problems of Biogeochemistry." Although George Vernadsky had translated both of these papers, Hutchinson would only publish the second of the two, and this heavily expurgated, in the *Transactions of the Connecticut Academy of Arts and Sciences* in June 1943. Hutchinson had thus begun a project of introducing an "expurgated" Vernadsky to the American public for the purposes of promoting his own genocidal agenda.

And what was Hutchinson's agenda? In December 1948, Hutchinson published a paper in *Scientific Monthly* entitled "On Living In the Biosphere." While he did not on this occasion try to drag in the name of Vernadsky, he clearly is starting to pave the way in that direction:

Looking at man from a strictly geochemical standpoint, his most striking character is that he demands so much—not merely thirty or forty elements for physiological activity, but nearly all the others for cultural activity… We find man scurrying about the planet looking for places where certain substances are abundant; then removing them elsewhere, often producing local artificial concentrations far greater than are known in nature. Modern man, then, is a very effective agent of zoogenous erosion, but the erosion is highly specific, affecting most powerfully arable soils, forests, accessible mineral deposits, and other parts of the biosphere which provide the things that *Homo sapiens* as a mammal and as an educatable social organism needs or thinks he needs. The process is continuously increasing in intensity, as populations expand and as the most easily eroded loci have added their quotas to the air, the garbage can, the city dump, and the sea.

Elsewhere in the same paper he writes:

The population of the world is increasing, its available resources are dwindling. Apart from the ordinary biological processes involved in producing population saturation already known to Malthus, the current disharmony is accentuated by the effect of medical science, which has decreased death rates without altering birth rates, and by modern wars, which one may suspect put greater drains on resources than on populations. Terrible as these conclusions must appear, they have to be faced.

---

5. See article by Meghan Rouillard, "A.I. Oparin: Fraud, Fallacy, or Both?" *21st Century Science & Technology*, Spring 2013.

The whole Russellite program is concisely presented in these remarks. To bring Vernadsky into this mix required some serious elisions in the written record.

In the paper published in the *Transactions*, Hutchinson eliminated entirely Vernadsky's first paper in the series "Problems of Biogeochemistry," on the pretext that "its propositions have become well-known through the other writings of the author (Vernadsky) and of his students, and there is no need of a translation at the present time." This was quite a remarkable statement given that almost none of Vernadsky's works had then (June 1944) been published in English.[6] Hutchinson readily admits with regard to the second paper that "some abridgement has been found desirable for the sake of clarity, but it is believed that all the ideas set forth in the original have been preserved in the present text." Fat chance that that will happen!

Vernadsky, who knew of Hutchinson through his work on limnology and through his son's and Petrunkevitch's letters, was excited by the fact that his paper would be published in the United States. He was following the project closely through correspondence with George and expressed gratitude to Hutchinson for taking it on. But getting the final copy, he was somewhat taken aback by some of the cuts made by Hutchinson. Writing to George on September 15, 1944, Vernadsky expressed his concern:

I'm very grateful to you and Hutchinson. I'm just not in agreement with the omission on page 502 of the reference to Dana [geologist James Dwight Dana], who established the empirical generalization of the role of the central nervous system in the course of geological time. The power of the central nervous system increased by leaps and bounds. You can observe this in any paleontology textbook.

It's funny that when I was working on this question in Moscow, I found at the Moscow University, after many years, American journals in which Dana defended himself against the theologians.

While Dana at a late stage in his career accepted the basic idea of evolution, he believed (unlike Darwin) that

Vernadsky's entire philosophical outlook was imbued with the knowledge that the mind of of man was a new and powerful geological force in the universe.

the process of evolution had a directionality to it, leading to the development of man and characterized biologically, as Vernadsky notes, by the development of the central nervous system. Dana, like Vernadsky, held that evolution had a directionality culminating in man in an epoch characterized chiefly by man's mental activity, which Vernadsky called the *noösphere* and Dana *cephalization*.

This was by no means the only cut that Hutchinson had made in the Vernadsky paper. He effectively eliminated almost all discussion of Vernadsky's seminal remarks on the work of Louis Pasteur on chirality and Vernadsky's idea of different "states of space."[7] Not unexpectedly, Hutchinson also eliminated portions of the manuscript in which Vernadsky expressed his unlimited confidence in the continuous progress of man's development through his creations of new ideas leading to technological advances. What remained was only a thin carcass of the real Vernadsky.

Soon afterward, in January 1945, Vernadsky's "Notes on the Noösphere" were published in *American Scientist* without such elisions. It is probable that George, who was sincerely intent on publishing his father's work in the United States and was aware of his father's concerns about the first translation, made sure that Hutchinson did not take a scalpel to this important statement. The "Notes on the Noösphere" also contains an extensive reference to the work of James Dwight Dana.

Creating a Green Movement

Of course, in 1944, it was an uphill climb in the United States, indeed, in the world at large, to introduce the notion of the genocidal population reduction program. The Second World War had done that all too effectively. "Population control" had been pretty much discredited by the Nazi program. And in the United States as elsewhere, there was a strong belief in the notion of scientific progress, similar to the belief so beautifully expressed in Vladimir Ivanovich's work at the time, specifically in his 1938 *Scientific Thought As A Planetary Phenomenon*. It would take a few decades before humanity would be prepared to accept these specious arguments in favor of its own demise.

The opportunity for introducing this "paradigm shift" in American society came in the 1960s. The brutal assassination of President John F. Kennedy and the initiation

6. The entire text of "Problems of Biogeochemistry, Part II" was published in English by *21st Century Science & Technology* (Winter 2000–2001). "Problems of Biogeochemistry, Part I" was also published by *21st Century Science & Technology* (Winter 2005–2006), utilizing the English manuscript copy of the text translated by George Vernadsky, and discovered in the Bakhmeteff Archives at Columbia University.

7. See article on Louis Pasteur, this issue, and Vladimir I. Vernadsky, "On the States of Physical Space" *21st Century Science & Technology*, Winter 2007–2008.

The brutal assassination of President John F. Kennedy was a decisive transformation of American culture away from its traditional notion of progress.

by President Johnson of that "long war in Asia" helped to plunge an entire generation of young people into a frantic search for "alternative life-styles." This was chiefly characterized by the hippie movement and its "back to nature" outlook. Here was an ideal opportunity to introduce on a broad scale those zero-growth ideas which had been anathema to an earlier generation.

In 1970, the "mainstream" scientific journal *Scientific American* devoted an entire issue to the theme of "The Biosphere." The introductory article was by none other than G. Evelyn Hutchinson. While he had not inserted Vernadsky's name in his 1947 diatribe, he would place it firmly in the center of this new effort to create a Green zero-growth movement. "The concept [of the biosphere] played little part in scientific thought," Hutchinson writes in his *Scientific American* piece, "until the publication, first in Russian in 1926 and later in French in 1929 (under the title *La Biosphère*), of two lectures by the Russian mineralogist Vladimir Ivanovitch Vernadsky. It is essentially Vernadsky's concept of the biosphere, developed about 50 years after [Eduard] Suess wrote, that we accept today."

The other articles in the magazine, dealing with the carbon cycle, the oxygen cycle, the nitrogen cycle, the role of agriculture, while written by different people, were also centered around the theme struck by Hutchinson: The activity of man on the planet is creating an ecological disaster and must therefore be limited.

Hutchinson, of course, could not completely eradicate Vernadsky's concept of the noösphere, so he simply asserted that Vernadsky had been mistaken in his view of human development. At the end of his article, Hutchinson writes:

Vernadsky, the founder of modern biogeochemistry, was a Russian liberal who grew up in the 19th century. Accepting the Russian Revolution, he did much of his work after 1917, although his numerous philosophic references were far from Marxist. Just before his death on January 6, 1945, he wrote his friend and former student Alexander Petrunkevitch: "I look forward with great optimism. I think that we undergo not only a his-

torical, but a planetary change as well. We live in a transition to the noösphere."

By noösphere, Vernadsky meant the envelope of mind that was to supersede the biosphere, the envelope of life. Unfortunately the quarter-century since those words were written has shown how mindless most of the changes wrought by man on the biosphere have been. Nonetheless, Vernadsky's transition in its deepest sense is the only alternative to man's cutting his life-time short by millions of years. The succeeding articles in this issue of *Scientific American* may contain useful hints as to how this alternative may be brought about.

Two years later, in 1972, a newly constituted Club of Rome issued a report called *The Limits To Growth*, which depicted an even more drastic scenario. The report was published by the UN Commission on Environment and Development. The Russellite agenda was thus introduced at the highest level of government. And now there was a mass movement of disenchanted youth around which to organize for this genocidal program.

And Vladimir Ivanovich Vernadsky was made into a guru of this new movement as well. New Age geologist and entrepreneur John Allen, who was spending his time in the early 1960s in San Francisco's hippie stronghold, Haight-Ashbury, with beat poet William Burroughs and others of his ilk, came across a book by Hutchinson entitled *The Ecological Theater and Evolutionary Play*, which also referenced the work of Vernadsky. Allen quickly placed Hutchinson's Vernadsky on the banner of a series of half-baked projects, beginning with a hippie commune in New Mexico, called Synergy Ranch, and later an upscale and alleged high-tech version of the commune, called Biosphere II, which he marketed as a predecessor to space colonization. Allen even succeeded in convincing some people from NASA, who had been bitten by the Green bug, as well as a number of otherwise serious scientists from Russia, that his up-scale hippy commune was the wave of the future in space exploration. Synergy Press also published the first

This Biosphere edition of the mainstream Scientific American was the first "shot across the bow" by the Greenie movement.

English translation of Vernadsky's *The Biosphere*—needless to say, in a heavily redacted edition.

James Lovelock, the so-called father of "climate change," with his thesis of Mother Earth, or Gaia, to whom mankind must bow in submission, also began to reference Vernadsky as his predecessor, even though he had no knowledge of Vernadsky before the 1980s.

As a result, to the extent Vladimir Ivanovich Vernadsky is known at all in the United States, he is widely seen in the form of Hutchinson's "ecological guru." *21st Century Science & Technology* and *Executive Intelligence Review*, both associated with the American statesman and economist Lyndon LaRouche, have taken upon themselves the task of introducing the real Vernadsky to the American public, to the American science community, and particularly, to the younger generation of Americans.

Vernadsky was one of the giants of science during the last century, a man whose ideas were often far ahead of his times. And science progresses by standing on the shoulders of its giants. Now when mankind is faced with the major scientific task of developing the new energy resources needed to support our growing population and of developing techniques here on Earth and in cosmic space for detecting and thwarting the threats that may face us from that region, as witnessed by the recent meteorite over Chelyabinsk, the thought—and spirit—of Vernadsky is more important than ever. By introducing the full depth of his scientific and philosophical achieve-

ments in the English language, we hope to provide American scientists with that giant, on whose shoulders they might stand from which to see a way forward for mankind, now enmired, in the worst financial crisis in history. Perhaps the optimism exhibited so strongly by Vernadsky, even in periods of repression and world war, may help to mobilize people today to begin to institute those needed changes which will enable mankind to launch a new era of growth and development in the "noösphere," and to help free a generation from that deadly mental illness known as "environmentalism."

Vernadsky Institute of Geochemistry and Analytical Chemistry

The Biogeochemical Laboratory founded by Vernadsky in 1929 now stands as the Vernadsky Institute of Geochemistry and Analytical Chemistry.

The Evolution of Species And Living Matter

Appendix to the French Translation of *The Biosphere*

by Vladimir I. Vernadsky

Translated from the French by Meghan Rouillard

Mollusk shells represent one of the first cases of biogenic migration of calcium.

From the Introduction to the French translation of *The Biosphere* written by Vladimir Vernadsky:

TRANSLATOR'S NOTE
The text is from a speech given by Vernadsky to the Society of Naturalists of Leningrad on February 5, 1928. All footnotes were written by the translator.

This book appeared in Russian in 1926. The French translation has been reviewed and in several instances, restructured, comparatively with the Russian text. It followed our essay on "Geochemistry," published in the same collection (1924), of which a Russian translation just appeared and of which a German translation will soon appear.

We will not give any bibliographic indications as they can be found in our "Geochemistry."

We have touched upon the same problems in various articles, of which the most important appeared in French in the Revue Générale des Sciences *(1922-1928) and in the* Bulletins de l'Académie des Sciences de Leningrad *(Petersburg) (1926-1927).*

The purpose of this book is to draw the attention of the naturalists, geologists, and above all, that of the biologists, to the importance of the quantitative study of life in its indissoluble relationships with the chemical phenomena of the planet.

We have tried to constantly stay on the empirical terrain without making hypotheses, a terrain which is still somewhat restricted, because of the small number of observations and precise, quantitative experiments which we had at our disposal.

It is essential at the present time to assemble in the shortest amount of time the greatest number of quantitative, empirical facts.

We cannot delay in trying to succeed in doing that, that is, only as soon as the great significance of the biosphere for living phenomena becomes clear.

Maybe this essay, whose purpose is to shed light upon this significance, will not go unnoticed.

I attach, as an appendix to the French translation, my speech, "The Evolution of Species and Living Matter," which seems to me to supplement the ideas established in The Biosphere.

Vladimir I. Vernadsky
December 1928

The Evolution of Species and Living Matter

1) Life constitutes an integral part of the mechanism of the biosphere. It is that which clearly stands out in the study of the geochemical history of the chemical elements: biogeochemical processes, so important, always require the intervention of life.

These biogeochemical manifestations of life constitute an ensemble of living processes, absolutely distinct, upon first view, from those studied by biology.

It still seems that there is an incompatibility between these two aspects of life, between its biological aspect and its geochemical aspect, and only a more profound analysis allows us to recognize the character of this difference.

It forces us to see that it is, in part, a question of identical phenomena which manifest themselves diversely, and in part, of living phenomena which are effectively different and considered differently; that is, either from the point of view of geochemistry, or, on the contrary, from that of biology.

The comparison of these two points of view transforms the scientific conception of the phenomena of life and gives more depth to it.

The difference between these two representations of life manifests itself in a particularly striking manner in the fact that the theory of evolution, which permeates the entire current biological conception of the universe, plays no role in geochemistry. Here, we will strive to shed light on the importance of the phenomena of the evolution of species in the mechanism of the biosphere.

From that standpoint, it is easy to convince oneself that the fundamental conceptions of biology must be submitted to radical modifications.

The species is habitually considered, in biology, from a *geometrical* point of view; the form—the *morphological characteristics*—are primary, in terms of importance. In biogeochemical phenomena, on the contrary, this is reserved to the number, and species is considered from an *arithmetic* point of view. Different species of animals and plants must be, in the manner of chemical and physical phenomena, composed of chemical compounds and physico-chemical systems, which are to be characterized and determined in geochemistry by *numerical constants*.

The morphological indicators which have been taken up by the biologists, and which are necessary for the determination of the species, are replaced by numerical constants.

In biogeochemical processes it is indispensable to take into consideration the following numerical constants: the mean weight of the organism, its mean *elementary chemical composition*, and its *mean geochemical energy*, that is to say the facility with which it produces displacements, otherwise called "the migration" of chemical elements in the living environment.

In biogeochemical processes, it is matter and energy which are primary instead of the inherent form of the species. The species can, from this point of view, be considered as a material analogous to the Earth's crust, as waters, minerals, and rocks, which, for the organisms, are the object of biogeochemical processes.

Seen from this angle, the species of the biologist can be envisaged as *living homogeneous matter*, characterized by mass, elementary chemical composition, and geochemical energy.

Normally, the characteristics of species are expressed by numbers informed by weight, chemical composition, and speeds of transmissions of geochemical energy, but these do not give anything but an abstract and very obscure idea of the reality.

It is possible to replace this idea with another, which relates more clearly to the character of the natural process which creates the organism. In this domain, we may take the point of view of physical chemistry and consider the organisms as autonomous fields where determined atoms in determined amounts are reunited.

This quantity constitutes precisely the characteristic

Drawings by Ernst Haeckel of early photosynthetic material.

The numbers obtained are very considerable: For example, concerning the *Lemna minor*,[1] the number of atoms for an organism is greater than 3.7×10^{20}, and reaches the hundreds of quintillions.

These great numbers correspond to reality, and lend themselves to numerical comparisons between the different species.

This determination of the species according to the number of atoms comprised in the volume occupied by the organism, only deals with the more customary biological characteristic of the species, which does not take into account the form and the structure.

The homogeneous living matter of geochemistry and the species of biology are identical, but the modes of expression are different.

2) The study of living phenomena in the mechanism of the biosphere shows differences which are still more essential, among the ordinary biological notions.

The biosphere in its fundamental traits has not changed, in the course of geological epochs, since the Archeozoic, since at least two billion years ago.

This structure reveals itself through a great number of corresponding phenomena, among them the biogeochemical phenomena.

Thus the geochemical cycles of the chemical elements seem to remain immutable in the course of geological time. The Cambrian should have the same character as the Quaternary Epoch or that of our days.

The conditions of climate, volcanic phenomena, and the chemical and physical phenomena of erosion have remained, in the course of all the geological epochs, as we now observe them. In the course of the entire existence of the Earth until the appearance of civilized humanity, not a single new mineral was created. The mineral species on our planet have remained the same, or were modified over time in an identical way. The same compounds as those of today have been formed for all time. In no case would we know how to relate a mineral species to a determined geological epoch. It is in this that the mineral species sharply distinguish themselves from living homogeneous matter, from species of living organisms. These latter modify themselves in a very marked way in the course of geological time; they spring forth and are always new, whereas the mineral species remain identical. Life, considered from the geochemical standpoint, as an element of the biosphere, submitted to simple

property of each organism and each species. It indicates the number of atoms that the organism of a given species can retain due to its force outside of the field of the biosphere, atoms which are drawn, thus, from the ambient environment. The volume of the organism and the number of atoms which compose it, expressed numerically, give the most abstract formula, but, at the same time, the most real measurement of how the species is reflected in the geological processes of the planet. We obtain this formula in measuring the size of the organism, its weight and chemical composition. This *number of atoms* and the volume of the organism thus determined are indubitably characteristics of the species. The presence of life in a sphere of determined volume with the concentration of a certain quantity of atoms constitutes a real phenomenon of nature, as characteristic for an organism as its form or physiological functions.

Fundamentally, this idea probably expresses with the greatest depth the essential traits of its existence.

_____
1. i.e., duckweed.

oscillations, taken in its entirety, appears as stable and immutable.

Life constitutes an integral part of the geochemical cycles which unceasingly renew themselves, but remain always identical, and it would not be able to undergo great changes in the course of phenomena studied by geochemistry. The mass of living matter, that is to say, the quantity of atoms captured by the innumerable autonomous fields of living matter, the mean chemical composition of the atoms of living fields, must, in sum, remain invariable across geological periods.[2] Moreover in the course of centuries, the forms of energy connected to life, the solar radiation and probably the atomic energy of radioactive matter, are not modified overall in terms of their amount.

We do not register in all these phenomena anything but oscillations, sometimes in one direction, sometimes in another, around a mean magnitude which appears to us as constant.

3) This immutability which characterizes all cosmic processes in the course of geological time, offers a striking contrast to the profound modifications undergone, at the same time, by the living forms studied in biology.

In particular, it is absolutely certain that all the characteristics of species, established by geochemical phenomena, are, again and again, radically modified throughout the geological epochs. Many a time have numerous animal and vegetable species disappeared, and new species were formed with a different weight, a different chemical composition, and another geochemical energy than those which preceded them. We cannot doubt that the chemical composition of bodies which are morphologically diverse is not altogether different. The extinct species necessarily corresponded to other forms of homogeneous living matter which have now gone extinct. Their numerical constants were different.

If, nevertheless, the general action of life remains identically the same in its details as compared, for example, with the phenomena of erosion, this indicates *the possibility of the formation of new groupings of the chemical elements, but not radical modifications of their composition and their quantity*. These new groupings do not affect

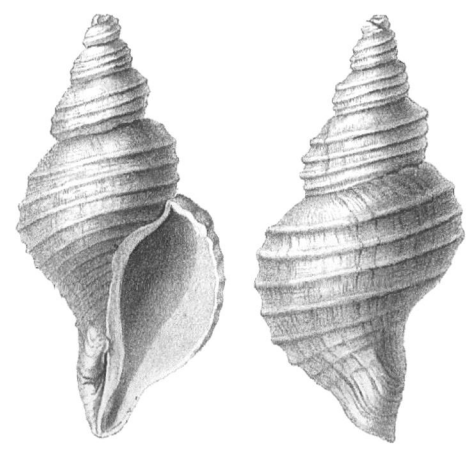

Mollusk shells mainly utilize calcium, whereas later structures like bone incorporate calcium in addition to other elements.

the constancy and immutability of geological processes.

It is a new fact of enormous importance for science, and we are beholden to its introduction into the domain of biology, to the geochemical study of life.

Whereas the morphological, geometrical aspect of life, taken in its entirety, undergoes great changes and continually manifests itself by the great evolution of living forms since the Archeozoic Era, the numerical, quantitative formula of life, always taken in its entirety, remains immutable in its essential proportions and, it seems well to be the case, in its essential functions.

It is true that the attentive study of the phenomena of evolution, in the case of biology, reveal the extreme irregularity of its progress. It cannot be a question of a constant change of all species, of all forms of life. On the contrary, certain species remain immutable for hundreds of millions of years, as, for example, the species of radiolaria from the Precambrian Epoch, which are impossible to distinguish from those of today; the same also goes for the species of the *Lingula*, which, since the Cambrian until our days, have not undergone a single change: they have stayed the same during the course of hundreds of millions of years, across innumerable generations which succeeded. We can cite a great number of analogous examples for periods which may not be as long, during which, if there were changes, they were, in any case, of little consideration. We can also, consequently, observe and study in living forms, not their *variability*, but their extraordinary *stability*. It could even be that this stability of forms of species over the course of millions of years, millions of generations, is the most characteristic trait of living forms, and merits the most profound attention of the biologist.

These purely biological phenomena are probably the manifestation of the immutability of life, considered in its essence in the course of all of geological history, immuta-

2. Ten years after making this speech, Vernadsky had altered his formulation either due to new data available to him or perhaps a more ontological reason: "*The mass of living matter of the biosphere is close to the limit and, evidently, remains a relatively constant value on the scale of historical time. It is determined, above all, by the radiant energy of the Sun, falling on the biosphere, and by the biogeochemical energy of the process of colonization of the planet. Evidently, the mass of living matter increases in the course of geological time, and the process of the occupation of the Earth's crust by living matter has not yet been completed.*" -1938, "On the Fundamental Material-Energetic Distinction Between Living and Nonliving Natural Bodies of the Biosphere."

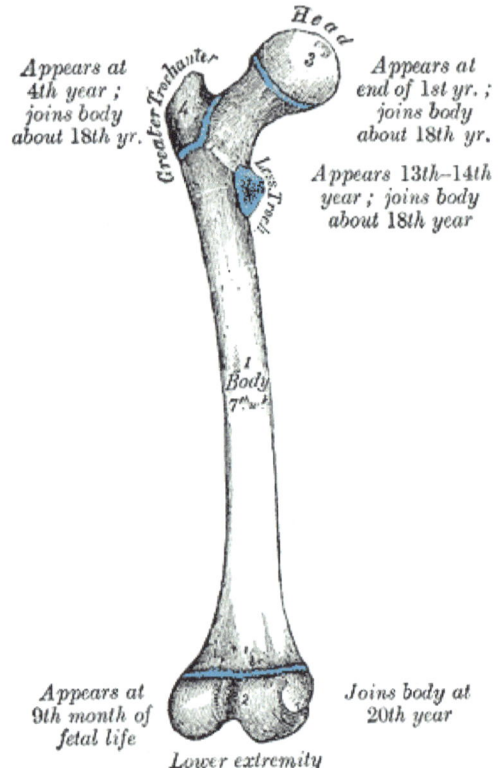

Appears at 4th year; joins body about 18th yr.

Greater Trochanter

Head

Appears at end of 1st yr.; joins body about 18th yr.

Appears 13th–14th year; joins body about 18th year

Less Troch.

Body 7th

Appears at 9th month of fetal life

Joins body at 20th year

Lower extremity

Bone incorporates other elements besides calcium, such as phosphorus.

bility which, in another form, is revealed through its role in the mechanism of the biosphere.

This stability of species, would merit, it seems, to draw more attention from the biologist than is currently the case.

The thought of the contemporary biologist orients itself in another direction. The evolution of forms in the course of geological time seems to be the most essential trait of the history of life; it embraces, for us, all of living nature.

This phenomenon was noted empirically, and in an absolutely rigorous way, one hundred years ago, by G. Cuvier, a naturalist of great profundity and precision, who demonstrated the existence of another universe, which we have ignored, of an earlier geological epoch.[3] This consideration served as a provocation during the lives of A. Wallace and C. Darwin, and later provoked a radical change of the entire conception of the scientific universe of the naturalist. The evolution of species occupies a central place in this conception, but draws attention to it to the point of forgetting about other biological phenomena which are just as important, if not more.

_____

3. This likely refers to Cuvier's theories on catastrophism. Here, we quote his 1796 paper on living and fossil elephants: "All of these facts, consistent among themselves, and not opposed by any report, seem to me to prove the existence of a world previous to ours, destroyed by some kind of catastrophe."

The notion of the evolution of species occupies a certain space in scientific thought such that a new phenomenon or a new explanation in the domain of biology must, to be admitted, relate itself to this in a more or less explicit way.

It is important to shed light on the manifestations of this evolution in biogeochemical processes, because the latest developments in geochemical studies are now stopped short for lack of facts which only the biologists can supply. The biogeochemical phenomena have to enter into the sphere of the biologists.

But, in addition, the research of the relationship which certainly exists between the evolution of species and biogeochemical phenomena is, on its own, of great interest.

This relationship between the evolution of species and the mechanism of the biosphere, and with the progress of living processes, is not in doubt. The fact that the essential numbers which characterize these processes are properties of species which modify themselves in the course of evolution, would suffice to prove it, and it is precisely the study of this relationship which will permit us to determine those relationships which exist between the immutability of the laws of life, considered in their entirety, in geochemistry, and its evolution, always considered in its entirety, in biology.

It is one of most important scientific problems of our time.

4) We can take on this problem starting from the study of the *biogenic migration of chemical elements of the biosphere*, characterized by the regularity of forms which it takes.

We will call the migration of chemical elements all displacement of chemical elements whatever may be the cause. The migration in the biosphere can be determined by chemical processes. For example, at the time of volcanic eruptions, it occurs by the movement of liquid, solid and gaseous masses. In the case of evaporations and of the formation of deposits, it is present in the movement of rivers, marine currents, winds, sediment transport and displacements of the Earth's crust.

The *biogenic migration* provoked by the intervention of life, thought of in its entirety, counts among the most grandiose, and also, typical processes of the biosphere, and constitutes the essential trait of its mechanism.

Innumerable quantities of atoms are submitted to the action of this uninterrupted biogenic migration.

It is not useful to insist, here, on the effect produced in the biosphere by a biogenic migration at a given scale. We have treated this question more than once.

It is important, nevertheless, to point out several essential traits of the biogenic migration, because it is indispensible to know them to understand what follows:

In the first place, *there exist several absolutely diverse forms of biogenic migration. On the one hand, the biogenic migration is linked in the most intimate way, and genetically, to the matter of the living organism, to its existence.* Cuvier gave a correct and precise definition of the living organism during its life, as an incessant current, a whirlpool of atoms which come from the exterior and return there. The organism lives as long as the current of atoms subsists. The current encompasses all of the material of the organism. Each organism on its own, or all organisms taken together, continually creates, by respiration, nutrition, internal metabolism, and reproduction, a biogenic current of atoms, which constructs and maintains living matter. In sum, it is the essential form and principle of the biogenic migration, of which the numerical importance is determined by the mass of living matter existing in a given moment on our planet. But this is not yet the entire biogenic migration.

Evidently, the effect of the entire biogenic migration does not depend directly on the mass of living matter. It does not depend any less on the quantity of atoms than on the intensity of their movements in intimate relation with life. The biogenic migration will be all the more intense as the atoms circulate more quickly; this migration can be very diverse, even while the quantity of atoms encompassed by life is identical.

That is *the second form of biogenic migration, in relation to the intensity of the biogenic current of atoms.*

There exists still a third. This third form begins to take on, in our epoch, the psychozoic epoch, an extraordinary importance in the history of our planet. It is the migration of atoms, also sustained by organisms, but which is not genetically or immediately related to the penetration or to the passage of atoms through their body. *This migration is provoked by technological activity.* It is, for example, determined by the work of burrowing animals, of which we notice traces since the most ancient geological epochs, by the consequences of the social life of building animals, termites, ants, and beavers. But this form of biogenic migration of chemical elements has taken on an extraordinary development since the appearance of *civilized humanity*, since tens of thou-

sands of years ago. Entirely new substances have been created in this way, as for example, metals in a free state. The face of the Earth transforms itself and virgin nature disappears.

This migration does not seem to be related directly to the mass of living matter; it is conditioned in its essential traits by the work of the thought of the conscious organism.

It is necessary finally, probably, in the fourth place, to also add the changes in the distribution of atoms provoked by the appearance, in the biosphere, of new compounds of organic origin. It is *probably, as for its effects, the most powerful form of biogenic migration.* It cannot yet be numerically evaluated, and I will not concern myself with it today.

This is the case, for example, for the migration which determines the release of free oxygen by chlorophyllic organisms, or that caused by the transformation of chemical compounds, unknown of until now in the biosphere, and for those transformations created by the genius of Man.

It is true that this type of chemical migration cannot always be easily distinguished from the first two. For example, the powerful chemical migration provoked by the destruction of bodies of dead organisms, is intimately linked to the processes of putrefaction and fermentation, sustained by the existence of special organisms.

Vernadsky points out that the "technological" mode of biogenic migration, as performed by burrowing animals, does not come close to what Man is able to achieve through his activity.

Photosynthesis from trees and forests allowed for a significant increase in the biogenic migration attributed to photosynthetic matter.

accomplished in the system will necessarily be expended. In the equilibria of this species, the work always reaches a maximum, whereas the energy in a free state tends towards a minimum.

Biogenic migration is one of the principle forms of work in natural systems of equilibrium and evidently it must tend towards a maximal manifestation.

We can consider this property of biogenic migration as an essential geochemical principle which governs, in an automatic way, biogeochemical phenomena.

The *first biogeochemical principle,* as I call it, can be formulated as follows:

"The biogenic migration of chemical elements in the biosphere tends towards its most complete manifestation."

But the biogeochemical processes do not explain this entirely.

5) The different forms of chemical migration indicated here constitute a special feature which we should have in view for the rest of our report.

Another characteristic trait is given by the physical laws which govern them.

Biogenic migration is only an element of another, still more powerful process in the biosphere, otherwise called the *general migration of its elements.* This migration is carried out in part under the influence of solar energy, of the force of gravitation, and the action of internal parts of the Earth's crust upon the biosphere.[4]

All these displacements of elements, whatever may be the cause, respond to diverse systems of equilibrium, which are mechanically determined; in particular, in the history of diverse chemical elements, they give birth to new, closed geochemical cycles, to whirlpools of atoms.

They can all be reduced to heterogeneous laws of equilibrium and to the principles formulated by Gibbs.

The cyclical processes in which the biogenic migration participates are maintained by an exterior force, renewed by an uninterrupted influx. The forces of radiant solar energy and atomic energy play a dominant role in the renewal of these elements.

These equilibria, studied outside of this exterior influx of energy, are mechanical systems, which necessarily arrive at a stable state. Their free energy will be zero at the end of the process, because all the work capable of being

6) Let us now examine how these two properties of biogenic migration manifest themselves in the biosphere: the first biogeochemical principle and the existence of the two forms of its manifestation—first, that connected to the mass of living matter, and secondly, to the technology of life.

The mass of living matter must, evidently, at the time of the maximum biogenic migration in the biosphere, reach the ultimate limits, that is, if there exist such limits.

The invariability of this mass seems to indicate that the biogenic migration of this form has more or less reached its limits since the earliest geological epochs.

This is not the case for the biogenic migration of elements which is related to the technology of life. Here we notice a sharp jump to our psychozoic geological epoch.

We aid in the development of this form of the biogenic migration and we must, in conformity with the first biogeochemical principle, admit that this form of the migration of elements will inevitably reach, with time, its maximum limit, while supposing that such a limit exists, or that it will constantly strive to reach its maximum development.

7) We can easily evaluate the correctness of the first biogeochemical principle in studying biogenic migration. The tendency which it has to attain its maximum development in the biosphere, can be observed in nature with respect to two phenomena: In the first place, the migration will occupy the greatest space possible, the maximum space accessible to it due to the mass of the living matter and the living technology inherent in this latter. The phenomenon manifests itself by the ubiquity of life in the bio-

---

4. Perhaps this refers to volcanism, or something similar.

sphere, as we see everywhere.

But biogenic migration, in that which concerns its geochemical action, not only depends on the quantity of atoms caught by it at every moment in the biosphere, but also on the rapidity of their movement, the number of atoms passing through living matter in a unit of time, or on the displacement, in this same unit of time, provoked by an intervention of a technological order by living matter within the ambient environment.

The first biogeochemical principle manifests itself, then, by the pressure of life,

Bird guano increased the biogenic migration of phosphorus.

which we effectively observe in the biosphere, and by the growing acceleration of the technological activity of civilized man.

It is especially important to take into account, at the same time, the phenomenon of the ubiquity of life, but also that of its pressure, and of the existence, in the biosphere, of living forms evolved in an environment of a radically different physical character.

We can and we must, fundamentally, admit that life manifests itself in two physically distinct spaces.

On the one hand, it appears in the gravitational field where we live. It is naturally the most customary for us.

But this gravitational field, where all is governed by the law of gravitation, does not embrace the entire domain of life.

The dimensions of the smallest organisms are akin to the dimensions of molecules, [although of another order of magnitude].[5] These organisms, whose diameter does not even attain one hundred thousandth of a centimeter, enter into the field of molecular forces, and their life, and their related phenomena, are not only regulated by universal gravitation, but are also submitted to the action of radiations which everywhere surround us: These radiations can overpower,[6] in that which concerns these organisms, the conditions of existence provided by gravitation.[7]

We know that these infinitely small organisms also enjoy this same ubiquity, fill the maximum space, and that the pressure of their life, the intensity of the current of atoms which they provoke, is extreme.

8) Thus, we can consider the ubiquity of life and its pressure, as an expression of the principle of ambient nature, which regulates the biogenic migration of chemical elements.

It is easy to convince oneself, when studying natural phenomena and the empirical facts which are treated therein, that the same ubiquity, along with the pressure of life, cannot be explained by the immutability of the present life of organisms.

These phenomena modify themselves in the course of geological time and develop, to a large extent, under the action of evolution.

The creation resulting from this evolution of new living forms, adapts itself to new forms of existence, augments the ubiquity of life, and enlarges its domain. Life penetrates, thus, the regions of the biosphere where it had not earlier had access.

We see, at the same time, how, in the course of geological epochs, new forms of life appear. Their occurrence leads, however, to an acceleration of atomic current through living matter, and also provokes, within atoms,

5. "*Bien qu'appartenant a une autre décade.*" The meaning of décade in this context is not clear.

6. The French word is *abolir*. It means to repeal or outlaw, but overpower seems more appropriate.

7. A footnote, to clarify and restate the idea expressed here, from *The Biosphere*: "...the field of stability of life is clearly divided into the field

of gravity for the more voluminous organisms, and the field of molecular force for the smaller organisms such as the microbes and ultramicrobes (on the order of 104 mm long). The lives and movements of the latter are primarily determined by luminous and other radiations. Although the size of these two fields are not well documented, we know that they must be determined by the tolerances of organisms."

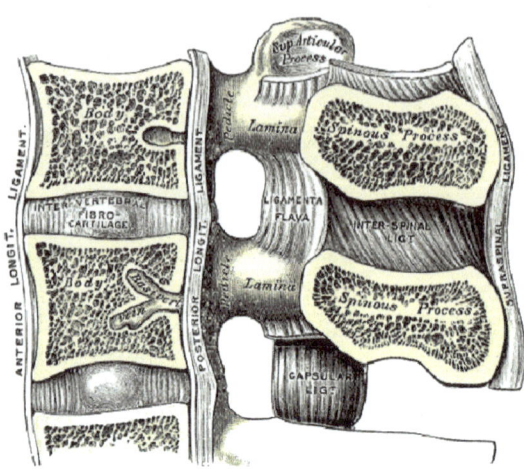

The intervertebral disk incorporates calcium fluoride, and in greater density than is found in other parts of the human skeletal structure.

new manifestations, unknown of until now, along with the appearance of new modes of displacement.

The attention already given by three generations of naturalists to the phenomena of the evolution of species has permitted the analysis of living nature, and convinces us that the ubiquity and the pressure of life observed everywhere, is radically modified and increased in the course of the geological epochs. *It is a result of the evolution and the adaptation of organisms to the environment.*

Two or three examples will suffice to make my thought more clear. The analysis of cave fauna show that it is composed of organisms having lived (at an earlier time) in the light. They adapted themselves to new conditions and thus enlarged the domain of life. This is also true for at least a portion of the benthos of the ocean. They adapted themselves to conditions of high pressure, cold, and darkness, while they originated from organisms having lived in other conditions.

It is a new phenomenon which enlarges the domain of life in the biosphere. The analysis of these phenomena also seems to indicate that the domain of life continues to enlarge itself in our geological epoch, also by the populating of the depths of the ocean.

In that which concerns other phenomena, we can still observe at each step identical processes. The flora and fauna of thermal sources, the flora and fauna of high altitudes or deserts, those of the glacial regions and those with perpetual snow, develop conforming to the laws of evolution. Life, in adapting to this in its environment, slowly annexes new regions, and reinforces the biogenic migration of atoms of the biosphere.

The process of evolution not only enlarged the domain of life, it intensified and accelerated the biogenic migration. The formation of the vertebrate skeleton, without a doubt, modified and augmented the migration of atoms of fluorine, in concentrating them, and the skeleton of aquatic invertebrates did the same for the migration of atoms of calcium.

It is not useful to insist upon the extreme increasing of the pressure of life in the biosphere caused by the appearance of the evolved *Homo sapiens*, who we can, it seems, name by combining the terminology of Linnaeus and that of Bergson in employing the three-fold characteristic of the species, "*Homo sapiens faber*" The thought of *Homo sapiens faber* is a new fact which fundamentally changes the structure of the biosphere after myriad centuries.

9) Thus, the analysis of living, ambient nature establishes in a sharp and decisive way that the ubiquity and the pressure of life in the biosphere are the results of evolution. Said otherwise, *the evolution of living forms in the course of geological time on our planet, augments the biogenic migration of the chemical elements in the biosphere.*

Naturally, the mechanical condition which determines the necessity of this character of atomic migration, is maintained uninterrupted in the course of all geological time and the evolution of forms has always taken this into account.

This mechanical condition which caused this biogenic migration of elements is due to the fact that life constitutes an integral part of the mechanism of the biosphere and, fundamentally, it is the force which determines its existence.

It is also evident that the evolution of species is correlated with the structure of the biosphere. Neither life, nor the evolution of its forms, would have been able to exist independently of the biosphere, nor to be divided from it as separated natural entities.

Starting from this fundamental principle, and the fact of the participation of evolution in the ubiquity and pressure of life in the current biosphere, we are well situated, concerning the evolution of living forms, to pose a new *biogeochemical principle.*

This biogeochemical principle which I will call the second biogeochemical principle can be formulated thus:

"The evolution of species, leading to the creation of new stable, living forms, must move in the direction of an increasing of the biogenic migration of atoms in the biosphere."

10) It is certain that this principle cannot in any way explain the evolution of species and does not enter into the tentative explanations of the different theories of evolution which now preoccupy the great thinkers. This principle regards evolution as an empirical fact, or rather as an empirical generalization, and attaches it to another empirical generalization, that of the *mechanism* of the biosphere.

But it is far from being indifferent, from the point of view of evolutionary theories, and it indicates, in my opinion, with an infallible logic, the existence of a determined direction, in the sense of how the processes of evolution must necessarily take place. This direction coincides perfectly, in its scientifically precise terminology, with the principles of mechanics, with all our knowledge of Earth's physical chemical processes to which biogenic migration strives.

All theories of evolution must take into consideration the existence of this determined direction of the process of evolution, which, with the subsequent developments in science, will be able to be numerically evaluated.

It seems impossible to me, for several reasons, to speak of evolutionary theories without taking into account the fundamental question of *the existence of a determined direction, invariable in the processes of evolution, in the course of all* the *geological epochs*.

Taken together, the annals of paleontology do not show the character of a chaotic upheaval, sometimes in one direction, sometimes in another, but of phenomena, for which the development is carried out in a determined manner, always in the same direction, in that of the increasing of consciousness, of thought, and of the creation of forms augmenting the action of life on the ambient environment.

The existence of a determined direction of the evolution of species can be precisely established by observation.

I will limit myself to a small number of general examples relative to the unfolding of processes of evolution, to paleontological indications considered from the point of view of the transformation of the biogenic migration in the course of geological epochs.

11) It was during the Cambrian period, at the limits of the ancient living world studied by us, when the higher invertebrates appeared. The fact in question is not absolutely established, but it is necessary to admit it to easily explain the sharp change which occurred shortly after the beginning of the Cambrian, concerning the conservation of organisms. The complete immutability in the course of the entire Pre-Cambrian period of the processes of erosion, their complete identity, if we consider their essential traits, with the analogous processes now, does not permit us to find the explanation of the absence of fossil remains in the different conditions of the surrounding environment.

There is not, at the same time, any reason to suppose that the metamorphosis of Earth's geological layers, occasioned by a determined duration of their processes, had, following this precise moment, an absence of organic fossils. It would otherwise be necessary to admit that all the oldest layers were completely transformed.

Now, we are quite familiar with the cases where the Pre-Cambrian layers were less metamorphised than those of the Cambrian and those of the more recent times.

It is probably the geologists, who here admit of a sharp

Photosynthesis caused a significant increase of the biogenic migration of carbon, oxygen, hydrogen, and nitrogen.

Feathers incorporate calcium, in addition to phosphosus. The activity of birds greatly increased the biogenic migration of phosphorus.

change of the *biogenic migration of atoms of calcium*, who are right. It is the first phenomenon of this type which we could establish.

It is possible that a similar modification of the biogenic migration of calcium, caused by the formation of new species endowed with skeletons rich in calcium carbonate, corresponds with the invasion of life into new domains of the biosphere. This modification must have had equal repercussions in the history of carbonic acid.

We can get an idea of the importance of this event by remembering the role, played in the biosphere, by the organisms which are very rich in calcium (the organisms containing it in fundamental preference to other metals), in the formation of the calcium deposits. The mechanism of the biogenic migration of calcium experienced great changes during the indicated time and this migration be-

came instantly more intense. In order to judge it by that which we know of the intensity of the migration of calcium, sparked off by the creation of the higher invertebrate skeleton, for example, that of mollusks or of coral, in relationship to those of the microscopic organisms, whose calcium is released, in the end, by water, it is necessary to admit an extreme and sudden augmentation of the intensity of its migration since the creation of these new forms of life.

At the beginning of Paleozoic life, and maybe at the Cambrian period, another very important fact relative to the biogenic migration of atoms calls itself to our attention: It is the radical transformation of the sylvan vegetation of the continents.[8] The process of gradual perfection of these organisms, of which the full blooming seemed to be attained, its point culminating in the Tertiary Epoch, still prolonged itself in the course of further geological epochs. This process corresponds to the conquest by life of a new and immense domain, that of the troposphere. The appearance of forests, exuberant with life, brought about a great change in the migration of atoms of oxygen, of carbon, of hydrogen, and simultaneously in that of all the living atoms of which the cyclical movement, first of all, had to become more intense, because the forests, especially the forests with leafy trees, persisting through new geological epochs, concentrated life, as much vegetable as animal, in proportions unknown of up until then. If we compare from this point of view the spore-bearing forests of primitive times to our tertiary forests of seed-bearing trees, the difference of the intensity of the biogenic migration will seem enormous to us.

During the Mesozoic Epoch, a new fact, the appearance of birds, augmented the intensity of the biogenic migration of atoms, and life again enlarged its domain. It was not until the Mesozoic Epoch and the Tertiary Epoch that flying organisms attained their fullest development, in the form of birds. Two very important biogeochemical functions attach themselves to these two new forms of life. We can hardly conclude that there is a relationship between these forms and the flying invertebrates which emerged very long ago, around the beginning of the Paleozoic, although these flying invertebrates, in particular, had fulfilled these functions and fulfill them still to this day. In any case, only the creation of the birds gave an impetus to the mechanism of the biosphere which it had not had earlier.

In the mechanism of the biosphere, in the biogenic migration of atoms, the birds, as well as the other flying or-

---

8. i.e., forests.

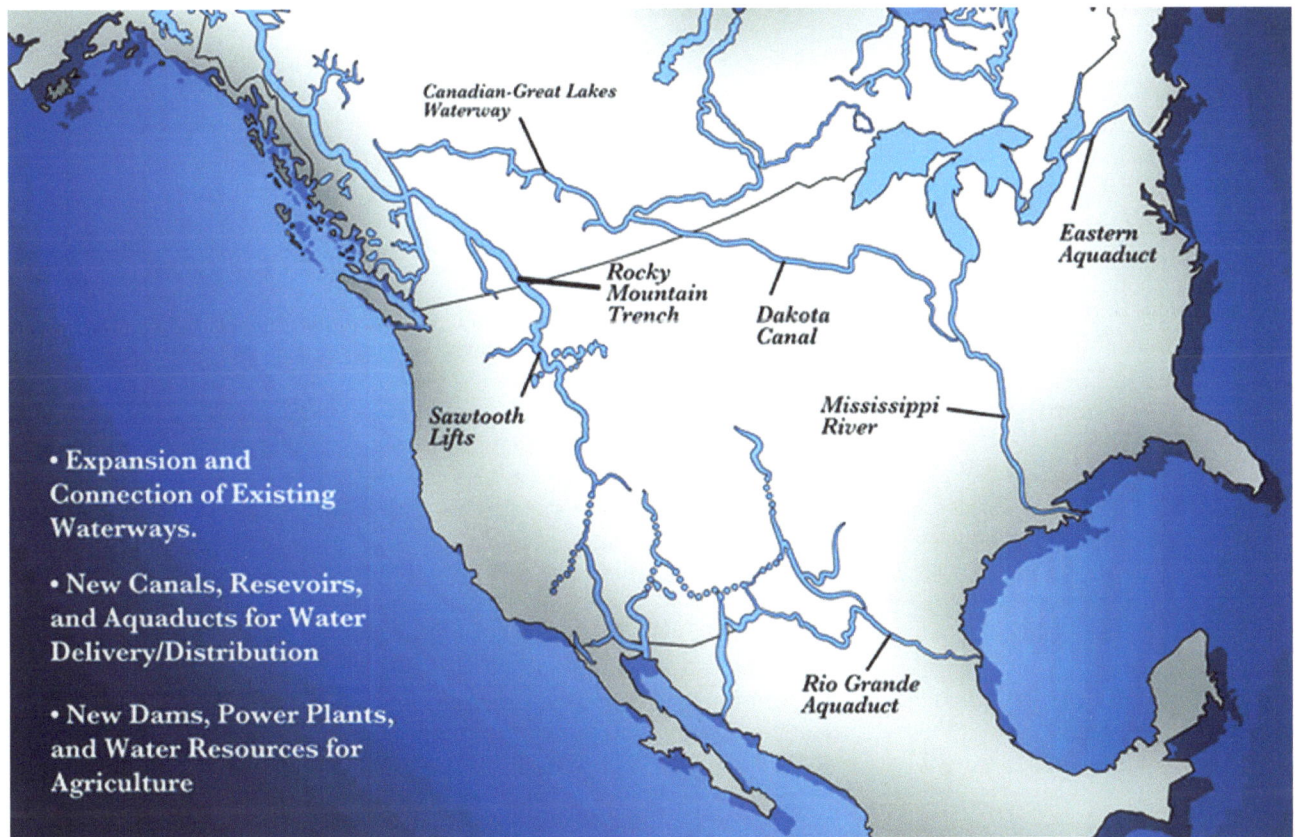

Man increases the biogenic migration of atoms in a way which is consistent with the increasing biogenic migration throughout evolutionary history, but on a scale not seen before in the biosphere. This would be achieved by building the North American Water and Power Alliance (NAWAPA) pictured here.

ganisms, play an immense role for the exchange of matter between the solid earth and the water, principally between the continent and the ocean![9] The role of the birds differs from that of the rivers, but as far as the quantity of mass transported, it comes close. The migrations of birds renders this role even more important in that which concerns the biogenic circulation of atoms. The appearance of these species of winged vertebrates not only created new forms of biogenic migration that affected the chemical balance of the sea and of the continent, but it also provoked a new wave of biogenic migration in the course of the history of discrete bodies, in particular, for that of phosphorus.[10] The winged invertebrates, the insects, did not play as important a role. It is true that the flying saurians (reptiles) appeared before the birds, but everything indicates that they did not exercise actions comparable to

theirs. The appearance of birds appears to be linked to that of new types of forests, or in any case seems to have coincided with them.

The role of civilized humanity, from the point of view of the biogenic migration of atoms, was infinitely more important than that played by the other vertebrates. Here, for the first time in the history of the Earth, the biogenic migration due to the development of the action of technology was able to have a greater significance than the biogenic migration determined by the mass of living matter. At the same time, the biogenic migrations changed for all of the elements. The process was rapidly effected in a relatively insignificant amount of time. The face of the Earth transformed itself in an unrecognizable way, and yet, it is clear that the era of this transformation has only just begun.

These transformations conform to the data of the second biogeochemical principle; the change led to an extreme growth of the intensity of the biogenic migration of atoms in the biosphere.

It is necessary to note here two phenomena: Firstly, Man (and this can not be doubted) is born of an evolution, and secondly, in observing the change which he produces

9. This exclamation point is in the French translation, but not in the Russian.

10. From Mark McMenamin's *Hypersea*: "The first aspect of Vernadsky's law involves actual transport by the motion of living organisms, such as the migration of phosphorus atoms from sea to land when shore birds leave their droppings inland. (To a Vernadskian, a seagull is phosphorus on the wing)."

in the biogenic migration of atoms, we note that it is a change of a new kind, which, with time, accelerates with an extraordinary rapidity.

We can then perfectly admit that the changes in the biogenic migration of atoms were effected in the course of paleontological periods under the influence of the creation of new animal and vegetable species, in a manner which is no less rapid.

The new quantitative form of the biogenic migration, corresponding to civilization, was prepared by the entirety of paleontological history. We would have been able to recover its first fossil remains, if we had known the laws of nature from the first pages of the annals of paleontology.

I stop myself here on several typical phenomena with respect to the evolution of species relative to the biogenic migration of the chemical elements. In all these cases, the agreement of evolution with the second biogeochemical principle is evident, as it always seems to manifest itself, in the analysis of the paleontological annals.

How did this agreement occur? Does it follow from a blind combination of circumstances or, indeed, from a more profound process, determined by the properties of life—incessant processes, always the same in their manifestations in the course of the entirety of the geological history of the planet? The future will decide this.

The regulating influence of the second geochemical principle will manifest itself in these two cases.

Even if the evolution of species happened randomly, accidentally, outside of the influence of the ambient environment, that is to say, the mechanism of the biosphere, a species which was accidentally created would, however, not have been able to survive and to enter into the whirlpool of the planet; at the same time, only the species which were sufficiently stable, and susceptible of augmenting the biogenic migration of the biosphere, would have survived.

It is, however, impossible to now oppose, in an elementary fashion, the organism in its environment, that is to say, the biosphere, as was done of old. We know that the organism is not an accidental inhabitant of the environment: It participates in the complicated mechanism and submits to fixed laws. Evolution itself constitutes a part of this mechanism.

The naturalist must exclude all the philosophical or religious notions, which have penetrated science from the outside, from his conception of the universe. For example, admitting the idea of the independence of the organism from its environment, and of an opposition between these two factors, would be a great error of this type.

From this point of view, there truly exists an intimate connection between the agreement of evolution and the principle which governs it and it is by no means a question of a simple confluence of circumstances.

12) Without preoccupying ourselves with the causes of evolution, while only indicating the necessity for it to have a determined direction, the study of biogeochemical phenomena thus circumscribes the domain of evolutionary theories admissible into science.

It seems that this study opens before us yet another domain of the phenomena of scientific activity, until now exclusively reserved to the speculations of philosophy or religion.

The new form of biogenic migration, at least new to this scale, was provoked, as we see, by the intervention of human reason.

However, it does not distinguish itself in any of the other manifestations of biogenic migration, which are connected to other vital functions.

We can, at the same time, establish in a precise way, that human thought changes in a sharp and radical way the course of natural processes, and modifies that which we call the laws of nature.

Consciousness, and thought, despite the efforts of generations of thinkers and wisemen, cannot be reduced to either energy or matter, however we define these bases of our scientific thought.

How can consciousness act on the work of natural processes which seem to be entirely reducible to energy and matter?

This question was last posed by the American mathematician A. Lotka, precisely on the question of biogeochemical phenomena. It is doubtful that his response was satisfactory. But he indicated the importance of the problem, and the possibility of tackling it.

It is probable that we will not be able to resolve this question until after having radically changed our fundamental physical notions, notions which have undergone and still undergo transformations with a rapidity for which we know of no prior examples in the history of thought.

The physical theories will inevitably have to concern themselves with the fundamental phenomena of life.

It is in this direction that thought now works, and it is impossible not to take into account these new and profound researches, among them the speculations of the mathematician and English thinker A. Whitehead; it is true, more philosophical than scientific, merit analysis. It is very possible that another English thinker, J. Haldane, was right in predicting, in the near future, a radical transformation of physics and its principles due to the introduction of the study of life phenomena into its sphere.

The study of biogeochemical phenomena, pushed to the forefront, allows us to precisely penetrate this domain of the connected manifestations of life and the structure of the physical universe, and, at the same time, future scientific theories.

This makes clear the profound philosophical interest which biogeochemical problems now present.

Biospheric Energy-Flux Density

by Benjamin Deniston

150 Years of Vernadsky

Tyrannosaurus Rex

I n a February 5, 1928 speech given to the Society of Naturalists of Leningrad, Vernadsky made a series of concrete arguments that go directly against the core ideology of what is generally recognized as the modern environmentalist, or "green" movement. The specifics of the argument made there have crucial implications for today.

One year later, Vernadsky included this speech in the 1929 French publication of his seminal work, *The Biosphere*. According to Vernadsky, "I attach, as an appendix to the French translation [of the Biosphere], my speech, 'The Evolution of Species and Living Matter,' which seems to me to supplement the ideas established in *The Biosphere*".[1]

Here Vernadsky directly addresses the evolution of life on Earth from the standpoint of his concepts of the distinct, but interacting phase-spaces of the biosphere, lithosphere, and noösphere, concluding that evolution has a direction, and a specific, irreversible form of progress:

This biogeochemical principle which I will call the second biogeochemical principle can be formulated thus:

The evolution of species, leading to the creation of new, stable, living forms, must move in the direction of an increasing of the biogenic migration of atoms in the biosphere...

[This second biogeochemical principle] indicates, in my opinion, with an infallible logic, the existence of a determined direction, in the sense of how the processes of evolution must necessarily take place... All theories of evolution must take into consideration the existence of this determined direction of the process of evolution, which, with the subsequent developments in science, will be able to be numerically evaluated. It seems impossible to me, for several reasons, to speak of evolutionary theories without taking into account the fundamental question of the existence of a determined direction, invariable in the processes of evolution, in the course of all the geological epochs. Taken together, the annals of paleontology do not show the character of a chaotic upheaval, sometimes in one direction, sometimes in another, but of phenomena, for

---

1. See Vernadsky's introduction to the French translation of *The Biosphere*. For an English translation of the introduction and the speech, see "The Evolution of Species And Living Matter," translated from French by Meghan Rouillard, *21st Century*, Spring-Summer 2012. All indented quotes below are from this translation of Vernadsky's speech on evolution.

which the development is carried out in a determined manner, always in the same direction, in that of the increasing of consciousness, of thought, and of the creation of forms augmenting the action of life on the ambient environment.

This concept runs in direct contradiction to the entire British reductionist school of thought which has increasingly dominated science over the course of the past century, and underlies the entirety of the founding of the modern environmentalist movement which has corrupted the morality of much of the population today. (See Box.)

This is not an academic debate: the governing beliefs in science and society have real-life consequences and effects. As Vernadsky clearly knew from his unique work on the concept of the noösphere, human progress can be studied in terms of the physical effects of scientific and cultural thoughts and discoveries. There are knowable benefits, or losses, resulting from either the successes, or failures, of humanity to progress scientifically and technologically.

For example, despite the depths of the immediate hyperinflationary crisis (the growing actual unemployment, the long-standing collapse of the productive capabilities of the trans-Atlantic region, unacceptable levels of global poverty and starvation, a looming collapse of food supplies for even developed nations, etc.), there is still a persisting delusion of investing in so-called "green jobs" and the "green economy," *activity inherently characterized by actually lowering the productive capabilities of the population per capita, necessarily resulting in mass death and suffering.*[2]

Understanding the principles underlying a scientifically definable nature of progress is of utmost importance for the immediate state of mankind. One path towards illustrating this principle is taking up Verandsky's challenge to identify the relationship between the overall progressive

_____

2. Mass genocide is not simply a consequence of the policy, but the explicit intention, as stated and demonstrated repeatedly by the founders and orchestrators of the green movement. See, "Behind London's War Drive: A Policy To Kill Billions," by Nancy Spannaus, *Executive Intelligence Review* magazine, November 18, 2011.

British Reductionism: Evolution from The Standpoint of Imperialism

One year after Vernadsky gave his Feb. 5, 1928 speech, and the same year as its French publication, British imperial establishment figures H. G. Wells and Julian Huxley published a book, *The Science of Life*, in which they reiterated the British school's rabidly reductionist view of evolution. This view was also clearly expressed by Alexander Oparin, with whom Vernadsky was directly at odds over the fundamental nature and origin of life. Wells and Huxley write:

Variation is at random, selection sifts and guides it, as nearly as possible into the direction prescribed by the particular conditions of the environment. Once we realize this, we must give up any idea that evolution is purposeful. It is full of apparent purpose; but this is apparent only; it is not real purpose. It is the result of purposeless and random variation sifted by purposeless and automatic selection. In brief, we are confronted with the gravest theological difficulties if we too light-heartedly set out to see purpose in Evolution. The wiser and saner course is to acknowledge our ignorance of ultimate causes and designs.

The concluding sentence of this quote is reminiscent of Adam Smith's famous quote from his *Theory of Moral Sentiments*:

To man is allotted a much humbler department... Nature has directed us to the greater part of these by original and immediate instincts. Hunger, thirst, the passion which unites the two sexes, the love of pleasure, and the dread of pain, prompt us to apply those means for their own sakes, and without any consideration of their tendency to those beneficent ends which the great Director of nature intended to produce by them.
—Adam Smith, *Theory of Moral Sentiments,* 1759

The view of Huxley and Wells sounds much closer to the stated social doctrine of the British Empire, imposed on populations for cultural and political effects, and less like valid scientific thought. This doctrine holds as a central axiom that there is no knowable purpose or inherent progress in the universe, and, if there were, mankind would have no business attempting to know the purpose, much less consciously intervene to determine his own fate. That is the doctrine the British Empire has fought to impose on the general population of the world in one form or another, be it in science, or in economics.

See Meghan Rouillard's article in this issue: "A.I. Oparin: Fraud, Fallacy, or Both?"

nature of the biosphere as a whole—as understood from his perspective of biogeochemistry and the role of living species—and the evolutionary change of living species within his understanding of the biosphere.

The evolution of species, leading to the creation of new stable, living forms, must move in the direction of an increasing of the biogenic migration of atoms in the biosphere... the agreement of evolution with [that principle] is evident, as it always seems to manifest itself, in the analysis of the paleontological annals.

How did this agreement occur? Does it follow from a blind combination of circumstances or, indeed, from a more profound process, determined by the properties of life-incessant processes, always the same in their manifestations in the course of the entirety of the geological history of the planet?

The future will decide this.

In honor of 150 years of Vernadsky, these questions are now re-assessed from the perspective of the 85 years of scientific work accomplished since Vernadsky delivered this speech.

The Conceptual Groundwork

Addressing Vernadsky's challenge will require drawing upon both the body of his life's work, but also the related discoveries of Lyndon LaRouche. LaRouche's founding of the physical science underlying the growth and development of human societies, *physical economics*, uniquely converges upon the same subject of study as Vernadsky's noösphere in a very important way.

Specifically, LaRouche developed the concept of *energy-flux density*, initially as an indispensable metric of economic progress. Measuring energy throughput, per unit time, per unit area, energy flux density proved to be one of the factors most intimately correlated with economic growth and progress.[3] Compare, for example, the vastly superior energy densities of nuclear reactions, fission and fusion, with chemical combustion (and especially with the ridiculously low energy density of wind and solar power systems).

Since demonstrating that *increasing* the energy-flux density of an economic system is critical to progress, LaRouche indicated that this characteristic could be generalized, as a property of other developing (anti-entropic) systems as well, such as in the development of life on Earth, or perhaps even in certain astrophysical processes. The "energy" measured will obviously be of a different form, but the general property of increasing density of action and change should remain as an indicator of progress.

As to Vernadsky's discoveries, while rejecting the fraud of Alexander Oparin, he fully promoted and continued the work of Louis Pasteur on what he referred to as the principle of Francesco Redi: *life only comes from life*.[4] In Vernadsky's work, living organisms not only express a distinct universal principle, but their domain of action, the biosphere, operates at a quantitatively and qualitatively higher rate of activity compared with that of the surface of a planetary body unaffected by life (the lithosphere). Thus the biosphere is superior, and has transformed the face of the planet at speeds and in ways impossible in the domain of the lithosphere devoid of life. Furthermore, the biosphere, driven by the evolutionary advance of life, has done this at successively higher and higher rates, defining a true direction and measure of progress in the history of life on Earth.

Moving beyond the biosphere, Vernadsky also recognized that human creative thought is a force absolutely distinct from anything expressed by simply animal life as such. The domain of action of scientific discoveries, of creative of human thought, identified as the *noösphere*, expresses a rate of activity growing much faster than that of the biosphere, overtaking and transforming the biosphere, raising it to higher rates of activity than the untended biosphere could ever accomplish.[5]

Overall, Vernadsky's revolutionary approach to evolution did not come from a foundation built on the characteristics of individual organisms, but rather sprang from his unique concept of the biosphere, and its tiered interaction with the lithosphere and noösphere. Vernadsky saw an overly narrow focus on individual species, abstracted from the context of the biosphere, as artificially limiting the investigation and thus preventing a fuller understanding of the nature of life.

It is also evident that the evolution of species is correlated with the structure of the biosphere. Neither life, nor the evolution of its forms, would have been able to

---

3. For an introduction into LaRouche's science of physical economics see his *So, You Wish to Learn All About Economics?*, New York: New Benjamin Franklin Publishing House, 1984.

Based on LaRouche's method, from the late 1970s through 1987 the economic staff of *Executive Intelligence Review* magazine (founded by LaRouche), produced a series of regular economic reports and forecasts which far surpassed official government and other private economic analyses produced over the same time period: the *EIR Quarterly Economic Report*.

4. See Vernadsky's three essays on the material-energetic distinction between life and non-life, "On Some Fundamental Problems of Biogeochemistry," "Problems of Biogeochemistry II," and "On the States of Physical Space." Available in the *21st Century* Winter 2005 (http://bit.ly/AxeuMd), Winter 2000 (http://bit.ly/wrL86T), and Winter 2007 (http://bit.ly/zYLPZY) issues.

5. For Lyndon LaRouche's analysis of the principled importance of Vernadsky's work from the standpoint of the historical continuity of the development of extended-European science with the history of science itself understood from the standpoint of physical economic progress in terms of the fundamental cultural development of human society. See LaRouche, "Vernadsky and Dirichlet's Principle", *EIR*, May 18, 2005.

exist independently of the biosphere, nor to be divided from it as separated natural entities.

This connection is intimately expressed in what Vernadsky identified as the *biogenic migration of atoms*, the continuous consumption, respiration, and other forms of material-energetic exchange between living organisms and the surrounding environment.

According to this understanding, living organisms become special kinds of singularities in the biosphere, composed of continuous fluxes of atoms, coming from, and returning to, the surrounding environment, but also, *more crucially, they are the energetic drivers of the entire biosphere, constantly shaping it, maintaining it, and bringing it to a more energy-dense and more highly organized state by their activity*. If living organisms were to stop their activity, the surface of the Earth would rapidly, in a geological "moment" of time, approach that of a planetary body like Mars.

Vernadsky shows, on this basis, that the evolution of living organisms is inseparable from, and the driving force in the development of the biosphere as a whole, while at the same time, an integral component *of* the biosphere, completely dependent upon it. Therefore, instead of solely focusing on the visible morphological structure of the organism, as is the practice of the standard biologist or paleontologist, the study of the material and energetic interaction of the living organism with its surroundings, the study of *biogeochemistry*, becomes absolutely indispensable in understanding the nature of the direction and progress in evolution.

From that standpoint, it is easy to convince oneself that the fundamental conceptions of biology must be submitted to radical modifications.

The species is habitually considered, in biology, from a *geometrical* point of view; the form—*the morphological characteristics*—are primary, in terms of importance. In biogeochemical phenomena, on the contrary, this is reserved to number, and species is considered from an *arithmetic* point of view....

In biogeochemical processes it is indispensable to take into consideration the following numerical constants: the mean weight of the organism, its mean *elementary chemical composition*, and its *mean geochemical energy*, that is to say the facility with which it produces displacements, otherwise called "the migration" of chemical elements in the living environment.

The current, abstracted view of a species, defined solely by its visual appearance (or by its DNA), while not useless, is not sufficient to define the history of life on the planet. What is needed is a study of the totality of a species, and of various species, their interactions, and their

ability to change and transform the surrounding environment. The action of the species in affecting the entire process of the biosphere becomes the primary point of reference, especially when that action is understood from its contribution towards creating a new, higher-order state of the biosphere.[6]

Vernadsky on Evolution

From this vantage point, Vernadsky converges on a measure of progress in evolution that falls under the concept of *energy-flux density* independently developed by LaRouche.

The ability of organisms or species to perform action in the biosphere Vernadsky called *geochemical energy*. In this way the displacement of chemical elements from one location to another, or from one form to another, by organisms can be measured. In his "Evolution of Species and Living Matter," Vernadsky focuses on three forms of this *biogenic migration of atoms*.[7]

1. The basic biogenic migration created by living organisms:

The living organism during its life, is an incessant current, a whirlpool of atoms which come from the exterior and return there. The organism lives as long as the current of atoms subsists. The current encompasses all of the material of the organism. Each organism on its own, or all organisms taken together, continually creates, by respiration, nutrition, internal metabolism, and reproduction, a biogenic current of atoms, which constructs and maintains living matter. In sum, it is the essential form and principle of the biogenic migration.

2. The rate or intensity of the biogenic migration of atoms:

Evidently, the effect of the entire biogenic migration does not depend directly on the mass of living matter. It does not depend any less on the quantity of atoms than on the intensity of their movements in intimate relation with life. The biogenic migration will be all the more intense as the atoms circulate more quickly; this migration can be very diverse, even while the quantity of at-

---

6. This is similar to a physical economic pedagogy of LaRouche. Taking a standard auto mechanic in the economy, can we really define the value of his actions, his productivity, solely by the actions he performs as such? Say he makes the exact same repair on the exact same car of two different individuals. By standard monetary economic accounting, both repairs would supposedly have the same value, the same hours of labor, parts, etc. However, if the first individual is then able, by aid of the mechanic's actions, to continue his work asset-stripping industrial firms, while the second is then able to continue his work producing tractors for farming production, the *physical economic* value, defined by the contribution of the worker to the productive capabilities of the entire economy, is drastically different.

7 Vernadsky also cites a fourth kind, but does not elaborate on it in detail in that location.

oms encompassed by life is identical. That is the second form of biogenic migration, in relation to the intensity of the biogenic current of atoms.

3. The biogenic migration due to technological developments:

The migration of atoms, also sustained by organisms, but which is not genetically or immediately related to the penetration or to the passage of atoms through their body. This migration is provoked by technological activity. It is, for example, determined by the work of burrowing animals, of which we notice traces since the most ancient geological epochs, by the consequences of the social life of building animals, termites, ants, and beavers.

These are three expressions of the geochemical energy of living organisms in the biosphere. The organism, or species, is understood, thus, not solely by its morphological structure, but by its power to effect change, specifically measured in terms of the growth and expansion of the biosphere over the lithosphere, as, for example, measured in these three forms of biogenic migration.

Focusing on evolution specifically from his understanding of the inseparable material-energetic interdependency between living organisms and the biosphere, Vernadsky formulates what he calls his *second principle of biogeochemistry* (different from his three types of biogenic migration).[8]

Naturally, the mechanical condition which determines the necessity of this character of atomic migration, is maintained uninterrupted in the course of all geological time and the evolution of forms has always taken this into account. This mechanical condition which caused this biogenic migration of elements is due to the fact that life constitutes an integral part of the mechanism of the biosphere and, fundamentally, it is the force which determines its existence. It is also evident that the evolution of species is correlated with the structure of the biosphere. Neither life, nor the evolution of its forms, would have been able to exist independently of the biosphere, nor to be divided from it as separated natural entities. Starting from this fundamental principle, and the fact of the participation of evolution in the ubiquity and pressure of life in the current biosphere, we are well situated, concerning the evolution of living forms, to pose a new biogeochemical principle. This

biogeochemical principle which I will call the second biogeochemical principle can be formulated thus:

The evolution of species, leading to the creation of new stable, living forms, must move in the direction of an increasing of the biogenic migration of atoms in the biosphere.

Vernadsky argues that even if the total mass of living matter were to remain constant,[9] over evolutionary time there will still be an increase in the *rate* of the biogenic migration, that is, increase in the biogenic flux, per mass of living matter, per unit of time. Or, in LaRouche's terms, an increasing energy-flux density of the biosphere.

According to Vernadsky, this should be the key characteristic of the directional progress of evolution. Since Vernadsky's accomplishments, decades of new evidence have accumulated, providing a new basis to conclusively demonstrate his concept of the nature of irreversible progress governing the development of life on Earth.

The New Evidence

Various proxies provide indications of the conditions of the biosphere during past periods, and when viewed in light of Vernadsky's concept of the second biogeochemical principle, can provide excellent support for his views on evolution.[10]

First, evidence of the geochemical energy of species from previous periods is sought. This is not directly measured in absolute terms; rather, various proxies are investigated, either from the geological and biogeochemical records, or from descendants or holdover species from previous periods. An estimate of the geochemical energy of different taxonomic *classes*, for example, as opposed to species, often proves more insightful, because this taxo-

---

8. Vernadsky described his first biogeochemical principle as "the pressure of life," specifically: *"the biogenic migration of chemical elements in the biosphere tends towards its most complete manifestation."* This is expressed, for example, in the tendency of life to expand into every location of the biosphere that it is technologically and energetically capable of occupying.

9. Although in this 1928 speech Vernadsky discusses a relatively fixed total mass of living matter over time, by 1938 he argues that the total mass has increased over evolutionary time. Given the fact that estimates of even the current living biomass vary, speculations on the total living biomass of previous geological periods will not be discussed here, and the evidence for changes in the rate of activity per unit mass will be investigated instead.

10. A significant amount of supporting evidence for Vernadsky's second biogeochemical principle is provided by a relative handful of studies from the past three decades. The work of the authors of these studies is of great significance for Vernadsky's concept, and, when understood from his science of biogeochemistry, provide additional support for a long-overdue fundamental revolution in the scientific understanding of the history of life. The fact that such a revolution has not already occurred, can only be attributed to the insistence on interpreting the evidence from within the accepted framework of ideological biases, typified by the continuing legacy of the British reductionist school, as expressed in, for example, Thomas Gould's unoriginal arguments attacking the concept of direction and progress in evolution. Science is often held back not by the quality of the evidence, but by the quality of the assumptions by which the evidence is interpreted. A revolution in the understanding of evolution will require the perspective of Vernadsky's biogeochemical analysis, and the independent work of both Vernadsky and LaRouche on the science of anti-entropic systems.

nomic level often specifies key characteristics which define the geochemical energy of an entire set of species.[11] Key proxies are found in indications of:

1. The metabolic rates of organisms.
2. The development of more energy intensive modes of life, such as actively pursuing and consuming other animals for food, predation.
3. The technological ability of organisms to freely move through the biosphere, expand their reach into new domains, and to alter the surrounding environment through their actions.

These are understood from their correspondence with Vernadsky's three forms of biogenic migration listed above.

Second, species tend to rise and fall in rough correspondence. This parallel turnover can be especially clear when examining the middle-level classifications (around the order/class taxonomic levels), pointing to the possibility that the correspondence indicates a characteristic geochemical energy associated with interacting and interdependent sets of species (or orders/classes, etc.). The progressive nature of the development of life is most clear from this perspective of sets of various interacting orders/classes defining large-scale orderings to the material-energetic structure of the biosphere across geological time (both in terms of the rate of biogenic migration, but also, possibly, in variations in the specific chemical elements in circulation, and in the biological and chemical structures formed by them). As will be seen below, there is remarkable evidence that specific biospheric sets define periods of relative stability across long periods of geological time, indicating a single level, or stage to the entire biospheric system.

Third, from one stage to the next, the geochemical energy of the biospheric system increases, specifically in terms of the biospheric energy-flux density, demonstrating Vernadsky's principle of progress in evolution.

_____

11. For example, the difference, especially in terms of geochemical energy, between two different species of mice is much less significant than the difference between a mouse (representing the mammalian class) and a lizard (representing the reptilian class). The standard taxonomic order, from low to high, is: species, genus (plural genera), family, order, class, phylum, kingdom.

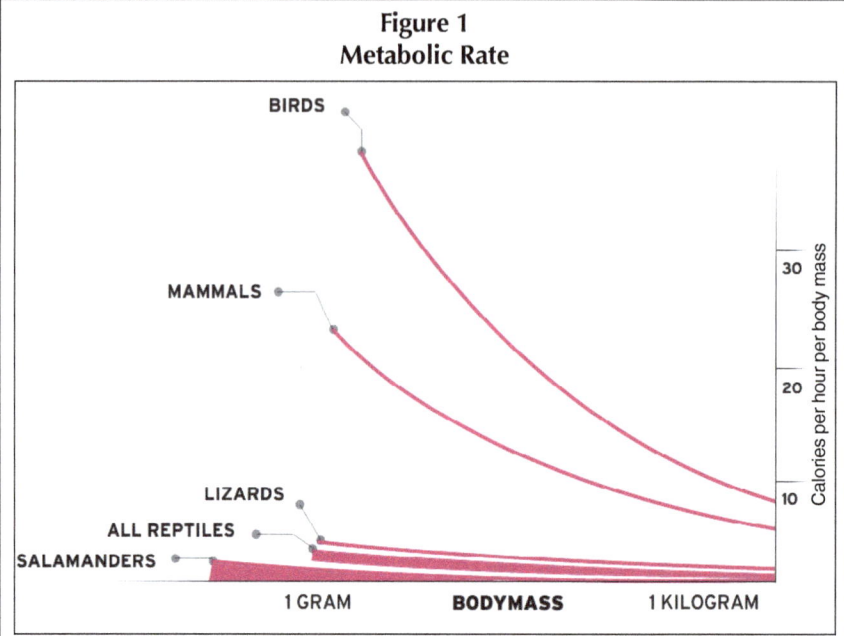

Figure 1
Metabolic Rate

Adapted from "Amphibians and Reptiles as Low-Energy Systems," by F. Harvey Pough, in *Behavioral Energetics: The Cost of Survival in Vertebrates*, Ohio State University Press, 1983.

For any given species, or certain classes, the metabolic rate of an organism will scale with the size of the organism in a specific, fixed way. However, different species or classes of organisms will have different values of the entire class, such that comparing examples of the animals of the same weight from different classes yields different metabolic rates. Here metabolic rates are expressed per unit mass of various classes and species of vertebrate tetrapods.

Metabolic Rates and Biogenic Migration of the Second Type

To start, compare the metabolic rates of different classes of vertebrate animals: for example, today's birds, mammals, reptiles, and amphibians. Their average metabolic rates show a clear succession (see Figure 1). A lower metabolic rate translates into lower respiration and consumption, and thus a lower geochemical energy (a lower rate of displacement of the surrounding material of the biosphere).

The question then is: how have the metabolic rates (and thus the geochemical energy) changed over the course of evolution?

On a larger scale, it has been known that the past 400 million years have been characterized by the succession of the age of the amphibians (lasting until 250 million years ago), to the age of the reptiles (lasting until 65 million years ago), to the age of the mammals (see Figure 2). However there are many intricacies (such as the question of whether dinosaurs were warm- or cold-blooded) which prevent a *direct* application of the metabolic rates of living reptiles and amphibians to some of their more famous ancient forerunners, although there are likely certain characteristic similarities.

The common amphibians of today are different from those dominant forms of seemingly amphibian-like vertebrates of 300 million years ago. The ones we find today did not exist then, and the skeletons of those we dig up from millions of years ago do not exist now... at least not most of them.

There are curious cases, however, often referred to as "*living fossils*," which help provide a critical glimpse into the ancient past. These are identified as species which emerged in the distant past, and have remained for a long period of time, having changed little for millions of years. There are only a very few such species of any given group, and they are often found in remote locations, either having been isolated from the main mode of the biospheric animal system, or having found new, minor roles in new biospheric stages.

For example, a handful of strange egg-laying species of mammals still inhabit regions of Australia and the surrounding islands. These are mammals of the *monotreme* order, and only two forms exist, the echidnas, of which there are only four species, and the platypus, with only one species. As of recent studies,[12] it is thought that the echidna and platypus have existed for over 110 million years, and that the currently living species are representatives of this distant time. Thus they are often referred to as "living fossils."

Another strange grouping of mammals separates itself, *marsupials*, distinguished by their pouches, used to raise their young, instead of the placenta of the more common *placental mammals* of the biosphere today.

Monotremes, marsupials, and placental mammals comprise a set of three very distinct forms of mammals which show distinct energetic differences. All are warm-blooded, but some are more so than others. The monotremes have average body temperatures of only about 90°F. Marsupials maintain a higher average body temperature, around 95°F, which is still below the average body

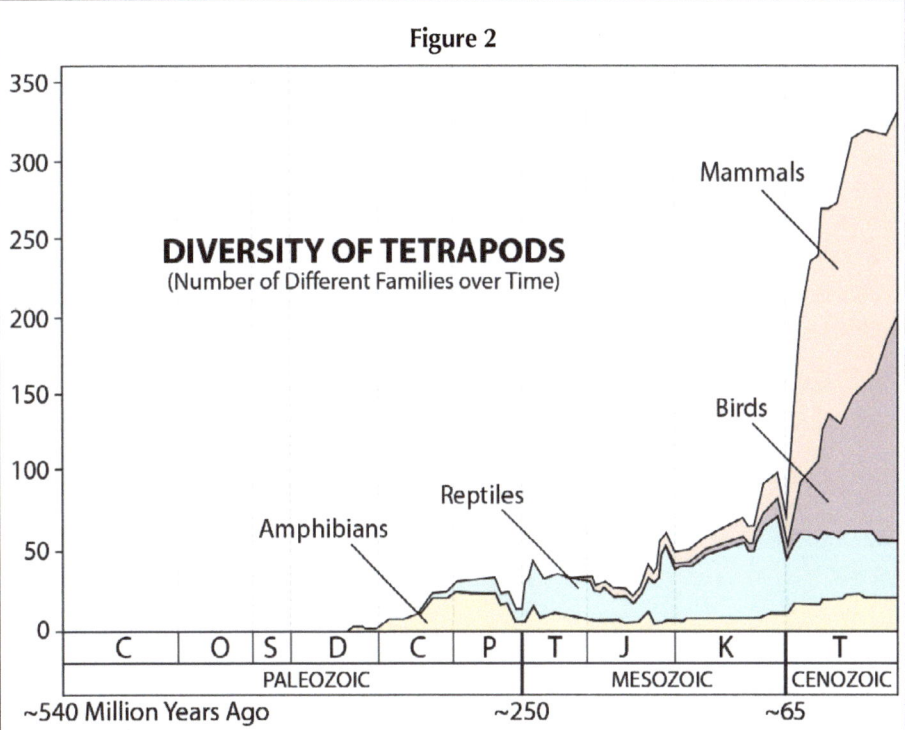

Figure 2

DIVERSITY OF TETRAPODS
(Number of Different Families over Time)

Generalized succession of dominant forms of vertebrates illustrated by the comparative number of known families over geological time. Examining the number of genera or species, instead of families, yields slightly different curves, but the same series of successions.

Image adapted from Michael Benton, "The history of life: large databases in palaeontology" in D. A. T. Harper (ed.), *Numerical Palaeobiology*. Wiley, Chichester, 1999, pp. 249-283.

temperature of most placental mammals, about 99°F. These different body temperatures correspond to the same succession of different metabolic rates, as indicated in Figure 3.

This indicates that the placental mammals express a characteristically higher geochemical energy. Based on Vernadsky's concept of progress being expressed in increased biogenic migration of atoms, this may be viewed as a fundamental reason for the global dominance of placental mammals (with an estimated well over 5,000 species), the lower role of marsupials (with a little over 300 species), and the tucked-away handful of monotremes (5 species).

Currently the latter two groups are mostly found in and around Australia, largely isolated from the core placental mammal mode. But that was not always the case. For tens of millions of years, South America maintained a strong and diverse marsupial population, including many species appearing remarkably similar to certain placental mammal parallels (such as a marsupial version of the saber-toothed cat, for example). This diverse marsupial population of South America flourished, as long as it remained separated from North America, which was the case for

12. "Molecules, morphology, and ecology indicate a recent, amphibious ancestry for echidnas," Matthew J. Phillipsa, et. al., *PNAS*, October 6, 2009, vol. 106, no. 40.

Figure 3

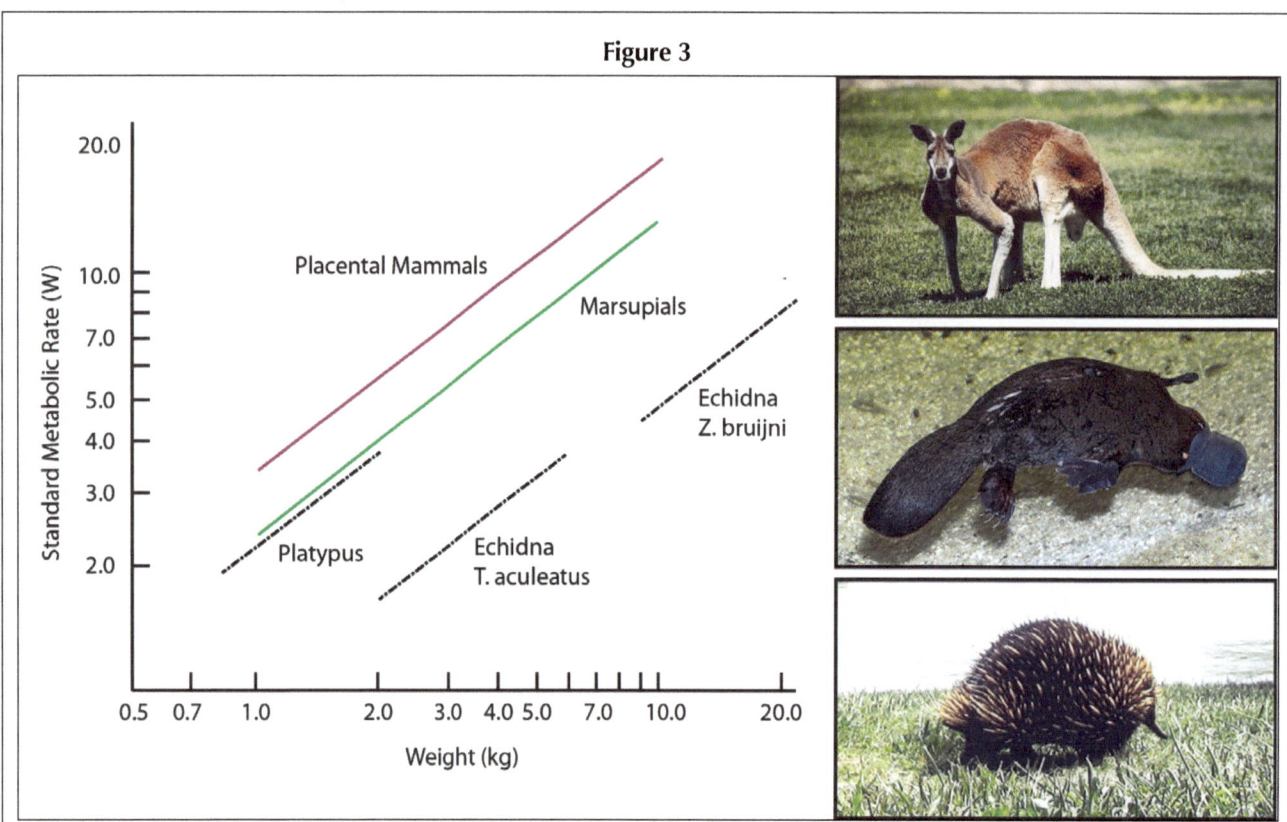

A graph showing the average metabolic rates of placental mammals, marsupials, and three species of monotremes (a platypus and two echidna) compared on a logarithmic scale. Different from the metabolic comparison in Figure 1, this measures the total metabolic rate of the whole organism (as opposed to the metabolic rate per unit mass). Pictures (top to bottom) of a kangaroo (marsupial), a platypus (monotreme), and an echidna (monotreme).

Graph adapted from page 144 of *Comparative Physiology: Primitive Mammals*, by Knut Schmidt-Nielsen, Carla Liana Bolis, and Charles Richard Taylor; Cambridge University Press, 1980. Echidna picture from wikipedia user Skyring, platypus picture credit Stefan Kraft. The adapted image is licensed under Creative Commons Attribution-Share Alike 3.0 Unported

tens of millions of years. About three million years ago, a landbridge re-connected South and North America (the formation of the Isthmus of Panama) and for the first time, the placental species of the north moved into South America, largely overtaking and replacing the marsupial system with the higher-order placental mammal system, leading to the extinction of the vast majority of the South American marsupial system.[13] Although this also gave the southern marsupials a chance to migrate north, only a few species, such as the opossum, were able to integrate into the placental mammal system, but no extinction of placental mammals as a consequence of marsupial migra-

tion is recorded. The introduction of the placental mammals into a marsupial system had a completely different effect than the introduction of the marsupials into a mammalian system.

Today, the last small foothold of the marsupial system is on and around the isolated continent of Australia.

Marsupials aren't the only strange creatures tucked away in that corner of the planet. A second example is the creature known as the tuatara. Although looking like a lizard, the tuatara is actually a significantly different holdover from 200 million years ago (a time well before the modern lizards of today emerged). Currently there are only two living species, isolated to New Zealand and some surrounding islands. From the perspective of Vernadsky's geochemical energy, what stands out is the significantly lower metabolic rate, with average body temperatures half to a quarter that of comparable modern lizards. Keeping with the pattern of lower-energy systems being driven out by higher-energy systems, currently the tuatara species are being threatened because rats have be-

13. As one geologist who is an expert in the region stated quite frankly, "If the Isthmus of Panama [the landbridge] was not there, the world would be very different today. All the animals of South America would be unique marsupials, like in Australia, very different to today because they would never have been invaded and overtaken by all the species that colonized from North America." See "How the Isthmus of Panama Changed the World," April 13th, 2011, Smithsonian Journeys blog. (http://www.smithsonianjourneys.org/blog/2011/04/13/how-the-isthmus-of-panama-changed-the-world/)

gun to populate the islands, and are threatening to overtake the tuatara which had hidden in their lower-energy haven for tens of millions of years.[14]

One last specific example shifts the discussion from the Australian region of the planet to oceans all over the globe. For hundreds of millions of years, the ocean floors were populated with forms of two shelled creatures called brachiopods. They are actually the most common animal fossils found from the Paleozoic era,[15] due to both their abundance and the fact that they fossilize well. Despite the fact that they appear visually similar to clams, they are actually quite different, and separated by 600 million years of evolution.[16] Both these ancient brachiopods and our modern day clams (and their bivalve class) are filter feeders, constantly circulating volumes of water through their bodies and playing an important role in the biogeochemistry of ocean systems. The evolutionary transition from the domination of brachiopods to bivalves is a well-studied case in paleontology for a few reasons, but, specifically for the argument here, it expresses another example of increased geochemical energy of the biosphere. Tests on modern bivalves indicate nearly ten times the metabolic rate of modern brachiopods, translating to a higher rate of circulation of the ocean water, migration of chemical elements, transformation of material, etc., and they are more effective at filtering food from the water. In the paper, "Seafood Through Time," paleontologist Richard Bambach discusses how this is associated with bivalves having greater capabilities in the biosphere, and why the modern brachiopods are relegated to the outskirts of ocean floor communities where food supplies are low:

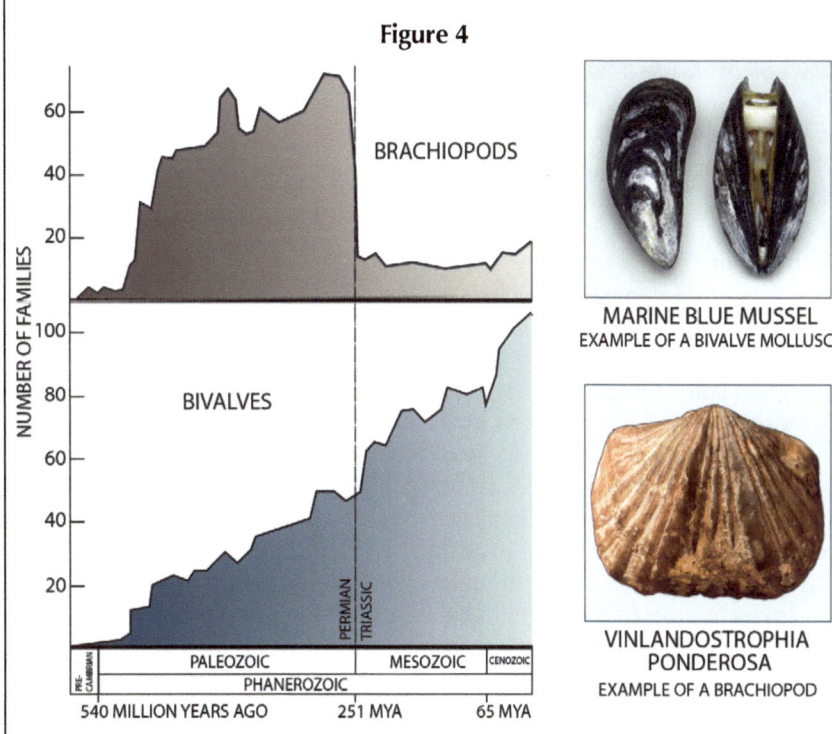

Figure 4

BRACHIOPODS

BIVALVES

NUMBER OF FAMILIES

PERMIAN
TRIASSIC

PRE-CAMBRIAN | PALEOZOIC | MESOZOIC | CENOZOIC
| PHANEROZOIC |
540 MILLION YEARS AGO   251 MYA   65 MYA

MARINE BLUE MUSSEL
EXAMPLE OF A BIVALVE MOLLUSC

VINLANDOSTROPHIA PONDEROSA
EXAMPLE OF A BRACHIOPOD

Comparison of brachiopods to bivalve molluscs over the past 540 million years in terms of the total number of families found at any one time.

"Clams and Brachiopods-Ships that Pass in the Night," *1980, by Stephen Jay Gould and C. Bradford Calloway; "Seafood through time: Changes in biomass, energetics, and productivity in the marine ecosystem," Richard Bambach; Paleobiology, Vol. 19, No. 3, Summer 1993, pp. 372-397. Blue mussel photo from wikipedia user Rainer Zenz.*

For example, Thayer (1981) called the sedentary, passive, suspension feeding articulate brachiopods "minimal organisms" and pointed out the variety of ways in which [brachiopods] function with low energy expenditure. In contrast... bivalve mollusks are more active. Many move around, even if sluggishly, some burrow actively, and some scallops can swim. The contrast extends to metabolic rates. Peck et al. (1989) reported that, for individuals of equivalent biomass under similar physical conditions, the oxygen consumption rate for the articulate brachiopod *Terebratulina retusa* (L.) is only 12% of that of the byssate bivalve *Mytilus edulis.* Thayer (1992) argues that the low energy requirements of articulate brachiopods accounts for their continued abundance in low food supply (oligotrophic) environments while bivalves dominate in more food-rich habitats.[17]

As expressed in the case of brachiopods versus bi-

---

14. Just recently, a New Zealand financier and rabid environmentalist, Gareth Morgan, has drawn international attention for promoting a campaign to eliminate all cats from New Zealand, including calls for possible mass euthanization of stray cats, because they are posing a threat to the native bird population of the island. The threatened birds include more unique holdovers, such as the New Zealand wattlebirds, of which there are only two remaining species, and likely stem from a split from other birds over 80 million years ago. New Zealand's famous oddity, the Kiwi, is also threatened by the mammalian invasion.

15. The Paleozoic era lasted from roughly 540–250 million years ago.

16. Clams (oysters, scallops, mussels, etc.) are part of the bivalve class of the mollusca phylum. Brachiopods makeup their own distinct phylum, based on fundamental structural differences.

17 Richard Bambach, "Seafood through time: Changes in biomass, energetics, and productivity in the marine ecosystem," *Paleobiology*, Vol. 19, No. 3, Summer 1993, pp. 372-397.

valves, and in other examples above, the general trend has been the displacement of less-energetic forms of life with more energetic ones. For example, in the oceans today, the only places that forms of life which used to dominate in Paleozoic era continue to reign, are in low-energy regions, with lower food supplies or on the fringes of the more productive regions of the biosphere.[18]

The examples above have only focused on comparing specific groups. An indication of the changes in the metabolic activity of the *entire ocean system* is provided by a 2002 study, examining tens of thousands of living and extinct genera over the past 500 million years.[19] This takes the investigation of geochemical energy out of specific examples, and begins to approach the entire biosphere.

The study took a total number of 40,859 distinct ocean genera recorded in the geological record and divided them into two groups: those characterized by higher metabolic rates and those characterized by lower rates.[20]

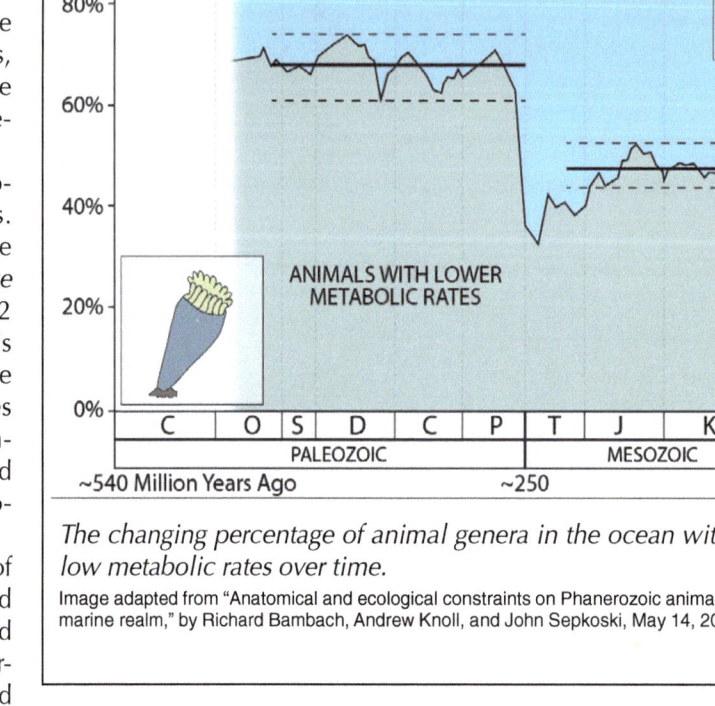

Figure 5

The changing percentage of animal genera in the ocean with high versus low metabolic rates over time.

Image adapted from "Anatomical and ecological constraints on Phanerozoic animal diversity in the marine realm," by Richard Bambach, Andrew Knoll, and John Sepkoski, May 14, 2002; *PNAS*.

They then examined the changing ratio between these two groups over geologic time. The results would grab Vernadsky's attention at once.

Two remarkable characteristics immediately jump out (see Figure 5). First, the relative stability of the biospheric system for many tens or even hundreds of millions of years: from about 445 to 250 million years ago, the division hovered around 70% of animals having lower metabolic rate, 30% higher. When a dramatic change occurred, ending this stage, the system re-stabilized at a new division of about 50 / 50. Approximately

65 million years ago, the last major shift brought the proportion of animals with lower metabolism to only about 35% and those with higher metabolism to 65%, thus nearly inverting the ratio from earlier conditions. At each stage, the values fluctuate around levels that are characteristic of that stage, suggesting that these levels are not accidental, but rather indicate a larger evolutionary structure of life, intimately tied to this concept of geochemical energy.

This analysis alone, examining the proportional structure of animal life in the oceans over time, provides very strong evidence for Vernadsky's second biogeochemical principle, illustrating progressive shifts in the internal ordering of the biosphere over time.

The evolution of species, leading to the creation of new stable, living forms, must move in the direction of an increasing of the biogenic migration of atoms in the biosphere.

This is life's increasing power to *change* the environment, doing so at successively higher rates. With each advancement, the lower order systems, such as the marsupials, dinosaurs, etc., are either discarded and replaced, or subsumed and transformed by the progression of life to a

18 "Seafood Through Time," Bambach, op. cit.

19 See Richard Bambach, Andrew Knoll, and John Sepkoski; "Anatomical and ecological constraints on Phanerozoic animal diversity in the marine realm," May 14, 2002; *PNAS*. Generally the fossil records from ocean life are more complete than comparable records of life on land, due to the better chances for organisms to be covered with sediment and preserved at the bottom of the ocean. This makes the marine fossil record a better subject for certain types of quantitative and qualitative analysis.

20 Their exact classification was slightly more complicated, but included the consideration of metabolic rates, which is being emphasized here. This was just one aspect of their study. See, Richard Bambach, Andrew Knoll, and John Sepkoski; "Anatomical and ecological constraints on Phanerozoic animal diversity in the marine realm," May 14, 2002; *PNAS*.

higher-order state.[21] Interestingly, a structure begins to emerge over the entire Phanerozoic eon, subsuming any one of the specific examples investigated so far. For example, the succession from the age of the amphibians, to the age of the reptiles, to the age of the mammals, defines the same three stages of activity as the changes in the percentage of ocean animals with higher metabolic rates.

This pattern continues to emerge when other examples of the increasing energy of the biosphere are examined.

The Case of Predation

Broadening the investigation beyond metabolic rates alone, other proxies indicate the dominance of Vernadsky's second biogeochemical principle.

For example, certain modes of life simply require more energy to maintain, such as predation. Whereas many ocean animals, especially of more ancient times, could survive by simply consuming organic matter from the seafloor, or by filtering food out of the ocean water as it flowed by, the action of actively pursuing another animal requires a more energetic mode of life. This is associated with higher metabolic requirements, but also an expanding food supply, technological developments, and a more energy-dense food web (higher biospheric capital intensity) to support higher level predators.[22]

When the fossil record is examined, it is revealed that over evolutionary time, predation has increased. This can

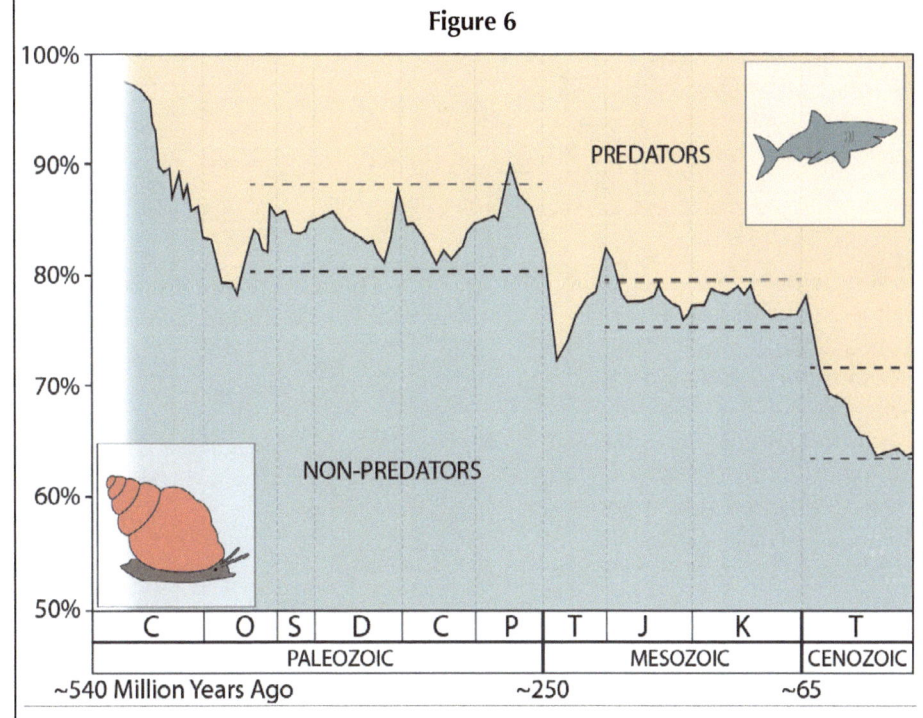

Figure 6

The changing percentage of animal genera in the ocean which are predators.

Image adapted from, "Anatomical and ecological constraints on Phanerozoic animal diversity in the marine realm," by Richard Bambach, Andrew Knoll, and John Sepkoski; May 14, 2002; *PNAS*.

be seen most clearly by taking the percentage of predator genera versus non-predator genera of the total known fossil population at any given time. Remarkably, although slightly less well-defined, the same general stages emerge as in the comparison of metabolic rates above. While there is some less-regular change between 540 to 445 million years ago, a roughly 200 million year period of relative stability occurs, in which the ratio fluctuates above and below the level of about 15% predators, 85% non-predators. This period if followed by a dramatic shift approximately 250 million years ago, again followed by a period of recovery, stabilizing around the level of about 23% predators, 77% non-predators, with some fluctuation above and below. The last major shift, although not appearing quite as clearly as in the previous analysis of this type, begins at about 65 million years ago, taking a bit longer to settle, but arriving at a level of about 35% predators, 65% non-predators.

These three levels of predation again indicate three successive stages of energy-flux density in the biosphere. So far, this investigation has generally focused on Vernadsky's second type of biogenic migration of atoms, the rate or intensity *directly due to organisms' consumption, respiration, etc.* Vernadsky's third type, the biogenic migration *due to technological developments*, also clearly expresses this development.

---

21 For example, numerous species of amphibians exist today, but the vast majority of the species are very different from those that existed 400 million years ago, and the role of today's amphibians in the mammalian stage is fundamentally different than their role in the amphibian stage.

22 See, for example, the 1993 paper by paleontologist Richard Bambach, "Seafood through time: Changes in biomass, energetics, and productivity in the marine ecosystem," *Paleobiology*, Vol. 19, No. 3, Summer 1993, pp 372-397. Bambach offers a series of arguments that are extremely valuable, and provide more conclusive evidence when viewed from the perspective of Vernadsky's concept of evolutionary progress. Examining an array of innovative proxies, Bambach presents a clear case for the increase of the total energy and energy density of life in the oceans over the past 500 million years. Predation is one example he focuses on. See "Seafood Through Time," Bambach, op. cit.

Biogenic Migration of the Third Type

A new series of proxies provide information about Vernadsky's third type of biogenic migration, which he defined as:

> The migration of atoms, also sustained by organisms, but which is not genetically or immediately related to the penetration or to the passage of atoms through their body. This migration is provoked by technological activity. It is, for example, determined by the work of burrowing animals, of which we notice traces since the most ancient geological epochs, by the consequences of the social life of building animals, termites, ants, and beavers.

This is expressed in a number of distinct ways. For instance, life has expanded into domains in which it did not exist prior, expanding the reach of the entire biogenic migration of atoms. Vernadsky gives the example of the development of birds, which now act as transporters of phosphorus and other chemical elements across the vast distances of their regular migrations. The movement of life onto land is another example, and perhaps the clearest: bringing the entire system of the biosphere to conquer and transform this new territory.[23] There are many useful examples within the ocean system as well.

On the ocean floors, the continental shelf area is generally the most populated with animal life. This includes animals which dig and burrow into the sediment of these shelf regions. The degree to which digging and burrowing animals have actively displaced and churned up the sea floor has increased over time. Going back to 540 million years ago, the records show that the average depth of life's displacement of the shelf floor was about 2-3 cm, with some sediments showing deeper, and others showing no disturbance. By 400 million years ago, 5-6 cm became

Figure 7

MOTILE ANIMALS ("FREE-SWIMMING")

NON-MOTILE ANIMALS (NOT "FREE-SWIMMING")

~540 Million Years Ago   ~250   ~65

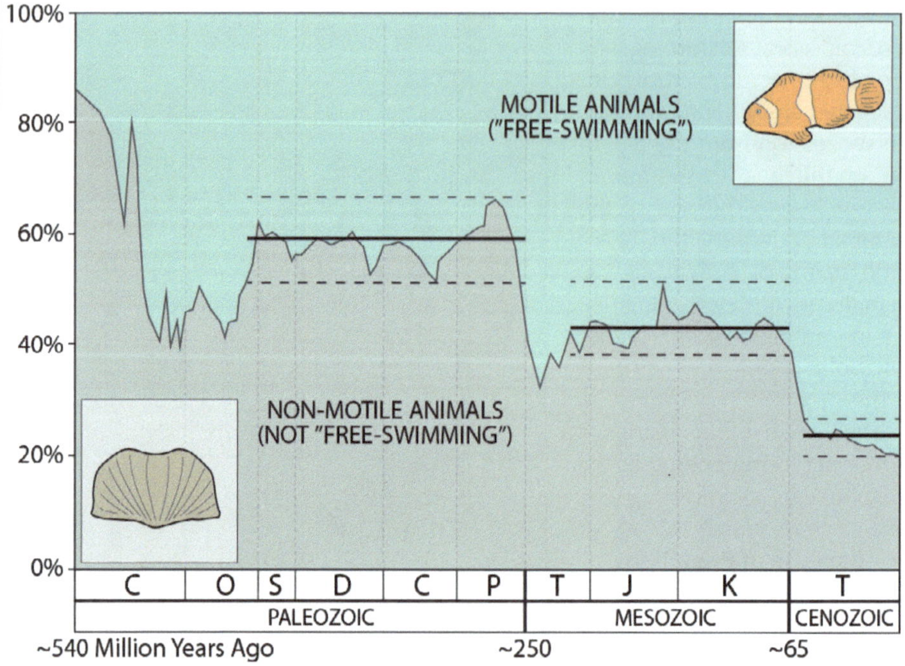

The changing percentage of animal genera that can freely swim and move around the ocean, compared with the percentage that are either stuck in one place, or which simply float with the ocean currents.

Image adapted from, "Anatomical and ecological constraints on Phanerozoic animal diversity in the marine realm," by Richard Bambach, Andrew Knoll, and John Sepkoski; May 14, 2002; *PNAS*.

average (again with some locations deeper and other locations with little or no disturbance). By 200 million years ago it became nearly impossible to find *any* layers of sediment, even as thin as 3-4 cm, that had not been disrupted by the activity of life. Starting about 65 million years ago, the activity has increased so much that certain forms of species that used to live by anchoring themselves in the sediment in earlier periods, could no longer live on the modern seafloor because the sediment was churned up and displaced at such a high rate they would be rapidly overturned, or even buried.[24] As Bambach stated in "Seafood Through Time:"

> Sediment disturbance by "biological bulldozers" (Thayer 1979) is now so severe that LaBarbara (1981) concluded that some reclining free-living bivalves which were abundant in the later Mesozoic, such as *Gryphaea* and *Exogyra*, would not be able to survive on the modern sea floor.

Thus, even something as simple as the ability of ani-

---

23 For example, the LaRouche PAC video, *The Hypersea Platform* (2011) presents the theory of the Hypersea, as developed by scientists Dianna and Mark McMenamin. http://larouchepac.com/hypersea-2011

24 "Seafood Through Time," Bambach, op. cit.

mals to move around freely has significant effects on the biosphere, expressing Vernadsky's third type of biogenic migration. To look again at the internal division within the biosphere, a similar comparison can be made between motile ocean animals that have the ability to swim freely around the ocean, and thus have the potential to participate actively in Vernadsky's third form of biogenic migration, versus those non-motile animals that do not, and are either stuck to one location, or simply float with ocean currents. Once more, the same pattern emerges when comparing the percentage of genera in the two categories:

• Some irregularity from 540 to about 445 million years ago.

• Beginning at around 445 million years ago, there is a distinct 200 million year time period when the ratio of self-moving to passive life fluctuates slightly above and below the average level of about 40% to 60%.

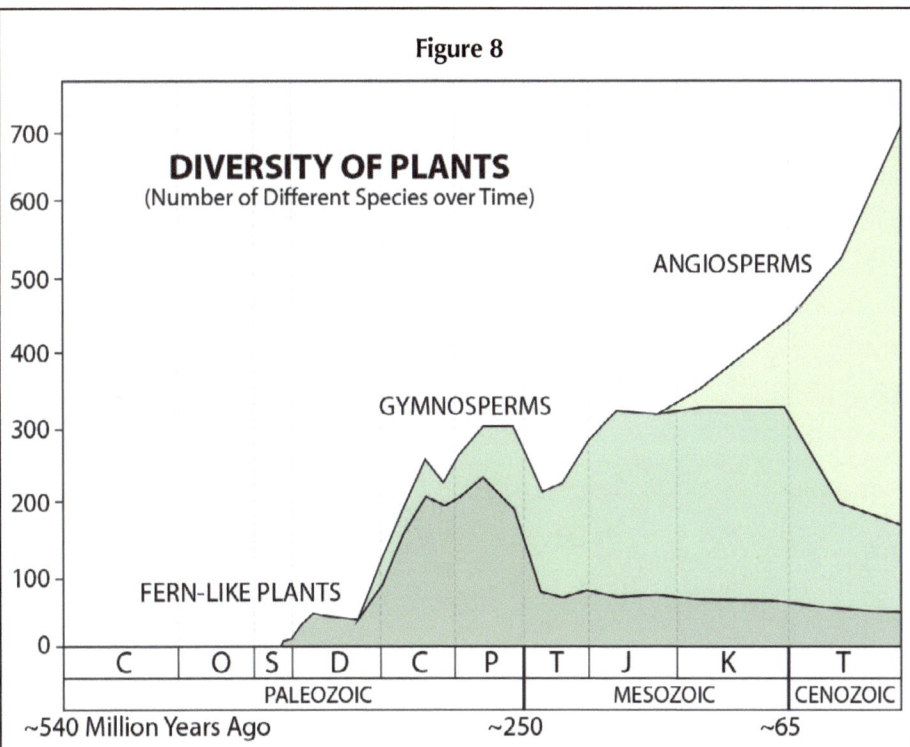

Figure 8

DIVERSITY OF PLANTS
(Number of Different Species over Time)

The biodiversity, counted in the number of different species, of three successive modes of plants, the pteridophytes (fern-like plants with spores instead of seeds), gymnosperms (with seeds but no flowers or fruit), and the angiosperms (flowering plants).

Adapted from, Niklas, Karl J. 1986. "Large-Scale Changes in Animal and Plant Terrestrial Communities." Pages 383-405 in D. M. Raup and D. Jablonski, eds., *Patterns and Processes in the History of Life.* New York: Springer-Verlag.

• About 250 million years ago, this changes and recovers to a new stabilized level of about 55% to 45%, with fluctuation above and below this average.

• The last major shift beginning about 65 million years ago resulted in a new proportion of the ocean animal system, with almost 80% now being motile, and 20% not.

These proxies taken together—the increase in metabolic activity, increasing percentage of predator species, greater expansion of animal life into new regions of the biosphere, the greater displacement of material of the biosphere, etc.—all indicate the general increase in the biogenic migration of atoms through the biosphere over time. Vernadsky's second biogeochemical principle is confirmed in each of his three forms of biogenic migration.

These studies, however, have only treated *animal* life thus far.

Because animal life depends upon the action of photosynthetic life, as well as other key components of the biosphere, the increase of the biogenic migration of the animal system should parallel changes in the photosynthetic activity and other characteristics of the biosphere as well.

Kingdoms United

Much has already been written about the significance of the motion of plant life onto land: bringing the water cycle onto land in a completely new way, transforming weathering and related activity; plants driving the transformation of rocky lands to nutrient rich soils; all of this activity feeding back into the oceans, providing nutrients and helping upgrade the ocean system as well.[25] The full significance of this process from the standpoint of Vernadsky's three forms of biogenic migration of atoms requires an entire study in itself.

Once firmly rooted on land, clear shifts in the dominant forms of plant life are apparent. The first mode is characterized by the initial domination of fern-like plants, which have spores rather than seeds, requiring wet or moist environments in order to reproduce. Approximately 250 million years ago, the first seed-bearing plants, the gymnosperms, which had emerged earlier as a minority, began to dominate. The development of the seed, with its

---

25 *The Hypersea Platform,* LaRouche PAC video, 2011, http://larouchepac.com/hypersea-2011.

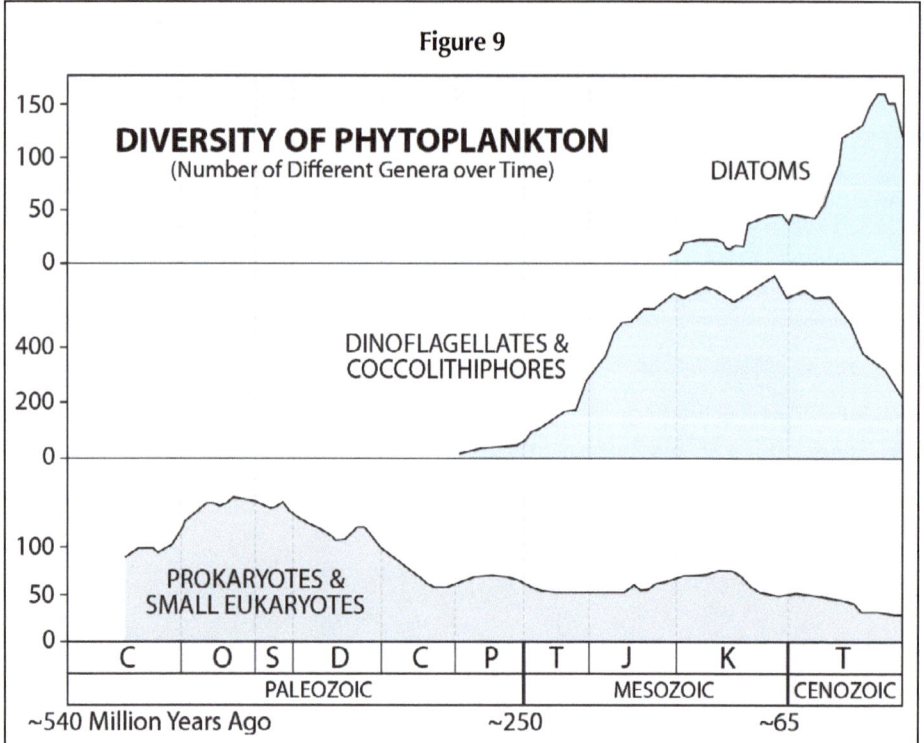

Figure 9

DIVERSITY OF PHYTOPLANKTON
(Number of Different Genera over Time)

DIATOMS

DINOFLAGELLATES &
COCCOLITHIPHORES

PROKARYOTES &
SMALL EUKARYOTES

| C | O | S | D | C | P | T | J | K | T |

PALEOZOIC · MESOZOIC · CENOZOIC

~540 Million Years Ago · ~250 · ~65

The changing biodiversity of the different modes of phytoplankton, measured in number of genera over time.

Adapted from "Evolutionary trajectories and biogeochemical impacts of marine eukaryotic phytoplankton," by Miriam Katz, et al., *Annual Review of Ecology, Evolution, and Systematics*, vol 35, 2004.

self-contained nutrient supply and internal fertilization, allowed plants to penetrate into drier regions of the land, forever changing the interiors of the continents. A third shift brought about a stage that is clearly associated with the shift in animal life around 65 million years ago, with the growing dominance of the flowering and nutrient-rich fruit-bearing plants, an increased energy density of sustenance which became crucial for the higher metabolic requirements of the mammalian system. Grasses (also flowering plants) emerged at this time as well, fast growing and providing the possibility for large grazing mammals.

In the oceans, the majority of photosynthetic activity is provided not by multi-cellular plants, but rather by tiny single-celled creatures called *phytoplankton*. Even in this separate kingdom, there are parallel shifts around the same stages, with one set dominating from 500 to 250 million years ago, transitioning to new types which dominated from 250 to 65 million years ago, and a third type coming into dominance around the shift 65 million years ago.

Changes in the biochemistry of these sets of phytoplankton coincide with shifts in the broader food web they support. The phytoplankton of the first stage (including cyanobacteria and other prokaryotes) rely more on the trace metal nutrients iron, zinc, and copper, while the phytoplankton of the second and third stages require higher proportions of manganese, cobalt, and cadmium. Paralleling changes in capital-intensity in a growing human economy, the increased diversity seen in the phytoplankton realm was outstripped by the increasing development in the animal world they support. The first group (the prokaryotes) could support around five species per single species of phytoplankton, while the second (dinoflagellates and coccolithophores) supported around 10, and the third (the diatoms) supported 60.

The phytoplankton introduced with the third stage, the *diatoms*, also uniquely brought silicon into the biogeochemical cycles of the oceans in a completely new way. Even more interesting, this directly paralleled the development of grasses, which were the first land plants to require silicon in significant quantities, and played a crucial role in converting silicon into a soluble form, and helped to deliver it in a usable form into the oceans, much to the joy of the diatoms. Together, this brought the silicon cycle under the control of the biosphere to a degree never before achieved. Diatoms had other technological developments which helped them achieve a new space in the biosphere: they acquired a better control over their nitrogen and carbon usage by developing a urea cycle. They also developed a unique storage vacuole that could hold excess nutrients, allowing a diatom to go through several divisions without needing external supplies. There is little doubt about the significance of the change in the photosynthetic baseline in the oceans for the entire animal system which depends upon it. As one scientist put it, "the [expansion] of the diatoms in the Cenozoic era demarcates a large change in the food-web structure of the Phanerozoic oceans."[26] As is clear from

---

26 The Cenozoic Era spans from about 65 million years ago to the present day, and the Phanerozoic Eon spans from the Cambrian explosion of about 540 million years ago to today. See, "Evolutionary trajectories and biogeochemical impacts of marine eukaryotic phytoplankton," by Miriam Katz, et al., *Annual Review of Ecology, Evolution, and Systematics*, vol 35, 2004.

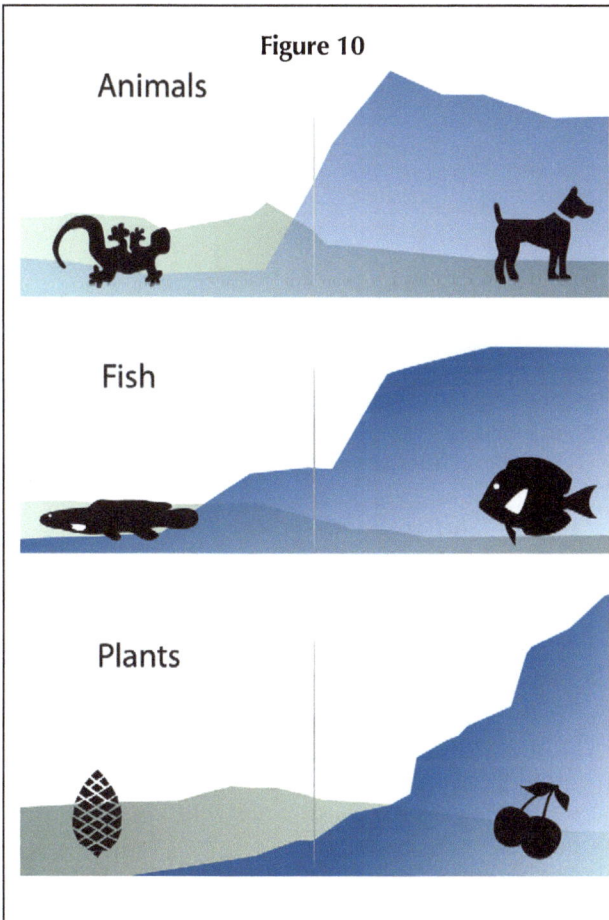

Figure 10

Animals

Fish

Plants

Phytoplankton

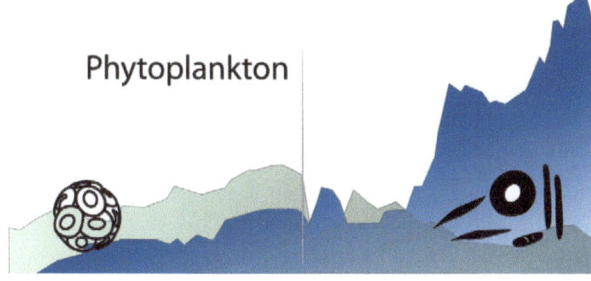

General biodiversity structure for different forms of life across the K-T extinction (marked here as the vertical line). Fish, tetrapods, land plants, and ocean phytoplankton (expressing the bounding top and bottom levels of the trophic system) all show the same character: the higher energy system starts in the second stage, grows in biodiversity, is less affected by the K-T extinction, and then comes into dominance. The lower order system is selectively more impacted by, or around, the K-T extinction, and declines thereafter, taking up a subsumed role within the new system.
Jablonski, eds., *Patterns and Processes in the History of Life*. New York: Springer-Verlag.

the above investigation, the change is towards higher rates of activity.

One last remark must be made about an often forgotten contributor to the biosphere, fungi. In Vernadsky's terms of the biogenic migration of atoms, fungi have played a consistent and important role, breaking down formerly living matter into forms that can be used by other living organisms.

Some biological structures decay more slowly than others. For example, lignin, a major structural component of plants, is one of the slowest plant structures to decay. It is also extremely hard to digest, and even animals that consume plant material do not consume lignin as a source of energy. If lignin is not consumed, or not broken down in some other way, it will just remain in that form for extended periods of time, being of little or no use to the biosphere, delaying the further useful biogenic migration of the chemical elements of which it is composed.

It was fungi that developed the capability to break down lignin, freeing carbon and oxygen, and increasing the rate of their cycling through the biosphere. Corresponding to the three stages of the biosphere discussed above, the development of this capability of fungi to break down lignin is part of the second stage, and associated with the overall increase of the biogenic migration of atoms associated with that stage.[27] Interestingly, fungi also go through an important shift associated with the third stage, the development of mushrooms, which are fruiting bodies with a higher density of consumable energy, allowing fungi to contribute to an increased food supply to the biosphere. All of this is well understood in terms of Vernadsky's second biogeochemical principle: the increase of the biogenic migration of atoms.[28]

Biospheric Energy Flux Density

From these examples from the plant and fungal kingdoms, the correspondence of developments within these kingdoms with development of the animal system is clear. Taken together, these become various proxies of a single metric, the general *biospheric energy-flux density* of successive geological periods.

The turnover from the popularized age of the amphibians, to the age of the reptiles, to the age of the mammals, defines the same biospheric shifts as the three stag-

---

27 "Seafood Through Time," Bambach, op. cit.

28 Interestingly, this development actually requires a net input of energy by the fungi. While fungi gain energy by consuming other plant material, for them to break down lignin they must lose energy to do so. It is as if they are working to contribute to accelerate the rate of biogenic migration of the whole biosphere, and not only acting in their own self-interest.

es of divisions of changes in the proportions of high versus low metabolic rates in the oceans. These shifts correspond with changes in other distinct sections of the biosphere: in photosynthetic life (both on land and in the oceans), in the percentage of species which are predators, in the ratio of animals which can freely move about the oceans, in the developments in fungal life, etc. These are all expressions of an overall increase in the

Toward A Physical Science of Anti-Entropic Evolution

For the large-scale development of the biosphere, the principle of progress is expressed clearly. From this understanding, the science of the anti-entropic system of the biosphere can be further refined, especially as mankind continues to dominate, manage, and increase the productivity of the biosphere, by applying the higher order creative power of mankind.

Addressing outstanding questions and challenges would further Vernadsky's hypothesis of the nature of the progress and development of the biosphere over time. This could provide a better understanding of the history of life, but also of how mankind can better manage and improve the biosphere (and also, eventually create a new biosphere on another planetary body). Such challenges include:

1. Refine possible metrics for biospheric energy-flux density. Something that can more specifically measured and/or estimated for both previous periods, and for current and future times.

2. Map the changes in the biogenic migration of atoms over the evolutionary development of the biosphere from the standpoint of the entire periodic table. How does the role of each chemical element evolve and change?[1] What about each isotope? How do the cycles and rates of key constituent elements of life, such as carbon or oxygen, change? What about trace metals and micronutrients, such as cadmium and copper, or the expansion of which elements and isotopes are used, such as the case of silicon? What can be learned about the changes and expansion of the different biological or biogenic structures formed in the biosphere? This could lead to an entire new periodic table, perhaps, and even provide for a better understanding of the distribution of natural resource deposits, or even the possibility of creating them biologically.

3. Investigate more specifically how the increasing biospheric energy-flux density corresponds to technological developments in the morphological structure of living organisms. For example, the development of wings, or a self-regulated body temperature. Many of these technological innovations made it possible for living organisms to expand into new regions which were simply out of prior reach.[2]

4. Develop a new taxonomic classification system from the standpoint of the process of the evolutionary development of the biosphere. As Vernadsky emphasized, solely defining the organism by its appearance is an abstraction. Instead, it must be investigated from the standpoint of its contribution to the entire process of the biosphere, and the evolutionary development of the biosphere over time.

5. Map out the times when living organisms fundamentally changed the material-energetic state of sections of the biosphere or lithosphere, making previously uninhabitable regions, habitable—the biospheric equivalent of infrastructure. Certain well-known case studies stand out, such as the motion of life onto land, or the oxygen revolution, but there may also be more subtle changes in the biogeochemistry or energetic conditions of the biosphere which have paved the way for new forms of life. For example, perhaps this could be seen in shifts in the utilization of the elements of the periodic table by the biosphere over time.

6. A study of the historical emergence and expansion of the noösphere in controlling and augmenting the biosphere. Recent studies have indicated that mankind's role is much stronger farther back in time than popular opinion has admitted thus far. Take, for example, the recent revelation that even the current nature of the Amazon rainforest is the product of mankind's activity going back thousands of years.[3] Plant and animal husbandry has gone back much farther. Irrigation systems have transformed deserts; fertilization has transformed soils. In terms of energy-flux density, productivity, and biogenic migration, what has this process looked like from the earliest times up to today, and how will we shape it into the future?

---

1. In his speech on evolution, Vernadsky gives a few examples, such as the changing role of calcium, or phosphorus. Meghan Rouillard discussed the dramatic changes in the biogenic migration of silicon in her video production, *Single-Celled Creativity*, http://larouchepac.com/node/17850

2. For example, in 1985 Richard Bambach produced an excellent analysis of most of the phyla and classes of the ocean fossil record over the past 540 million years, identifying specific technological shifts in the morphology of various species which were directly associated with the expansion of the biodiversity of that group. See, "Classes and adaptive variety: The ecological diversification in marine faunas through the Phanerozoic," by Richard Bambach, published in the book, *Phanerozoic Diversity Patterns*, 1985, Princeton University Press, edited by James Valentine.

3. See "Virginity Lost" by Fred Pearce, in the January-March, 2007, issue of *Conservation*.

biospheric energy-flux density of the entire system of life on Earth.

Provocatively, these transitions from one stage to the next are demarcated by the largest mass extinctions of the entire Phanerozoic eon (the last 540 million years). The initiation of the first stage is marked by the Ordovician-Silurian mass extinction of 445 million years ago, the second largest mass extinction of the eon. The division between the first stage and the second (250 million years ago) is provided by the largest mass extinction of the eon, the Permian-Triassic mass extinction, and the shift from the second to the third stage is demarcated by the famous K-T mass extinction which eliminated the dinosaurs, and approximately 75% of all species on the planet, about 65 million years ago.

Even more interesting, each of these extinctions is selective with respect to the organisms associated with the respective stages of biospheric energy-flux density. The biodiversity of the diatoms was hardly affected by the K-T mass extinction, whereas the phytoplankton of the second stage were much more severely hit, and never regained the diversity they had prior. Although mammals were affected, the K-T extinction was much more devastating to the reptilian class. The case is similar for angiosperms compared with gymnosperms, etc. (see Figure 10). The same character of energy-flux density selective extinction is clear in the transition from the first stage to the second, as seen in the comparisons of bivalves and brachiopods, reptiles and amphibians, fern-like plants and gymnosperms, etc. This is also reflected in each of the three charts showing the increasing percentages of animals with higher metabolism, which are predators, and are freely-moving, with each shift occurring at these mass extinctions.

Understood in this context, the traditional view of the mass extinction needs to be inverted. Instead of a self-defined event, interrupting some balance of life, the mass extinctions become merely shadows or effects of a primary process of anti-entropic growth in the biosphere. While there may have been particular triggers, such as an asteroid impact, which might have helped to affect the exact timing, or kick-start a transition, they are of secondary importance, and not the *cause* of the anti-entropic development of the biosphere.

As Vernadsky said,

Taken together, the annals of paleontology do not show the character of a chaotic upheaval, sometimes in one direction, sometimes in another, but of phenomena, for which the development is carried out in a determined manner, always in the same direction, in that of the increasing of consciousness, of thought, and of the creation of forms augmenting the action of life on the ambient environment.

These three stages of the development of life characterize the development of the biosphere in broad, but crucial strokes (see the table, *Three Stages of the Phanerozoic Biosphere*, in the appendix). Although much more work must be done to bring the investigation to greater resolution (see box), this should stand as undeniable proof of Vernadsky's great second principle of biogeochemistry, a principle of progress.

Willfully Acting on Principle

What does all of this mean for mankind today? Is mankind destined to be just another animal species, overtaken by the evolutionary process of the biosphere? Or perhaps by changes and developments in our Solar System, or even within our galaxy, which continue to have direct impact on the conditions of life here on Earth? This points to a more fundamental issue.

Progress is a principle of the universe in which we live. We certainly do not know the universe in its entirety, but we can know and understand this characteristic. Forms of existence that do not progress go extinct. Progress, per se, is not a vague, indefinable notion, but has a very specific character. For human society, this is expressed by a power that is completely absent from any form of animal life alone. It is the unique potential to act *willfully*, to creative new stages of nature, states which never before existed, and would be impossible for simply life, unaided by human creative action, to ever generate—in short, to willfully act in, and create our own future.

As Vernadsky recognized, this raises interesting questions. The action of human society can be seen in the unique quantitative and qualitative increase in the biogenic migration of atoms as a consequence of human activity.

The role of civilized humanity, from the point of view of the biogenic migration of atoms, was infinitely more important than that played by the other vertebrates. Here, for the first time in the history of the Earth, the biogenic migration due to the development of the action of technology was able to have a greater significance than the biogenic migration determined by the mass of living matter.

At the same time, the biogenic migrations changed for all of the elements. The process was rapidly effected in a relatively insignificant amount of time. The face of the Earth transformed itself in an unrecognizable way, and yet, it is clear that the era of this transformation has only just begun.

These transformations conform to the data of the second biogeochemical principle; the change led to an extreme growth of the intensity of the biogenic migration of atoms in the biosphere.

It is necessary to note here two phenomena: Firstly,

Man (and this can not be doubted) is born of an evolution, and secondly, in observing the change which he produces in the biogenic migration of atoms, we note that *it is a change of a new kind, which, with time, accelerates with an extraordinary rapidity.*[29]

These changes in biogenic migration, while completely unique to only human action, can be measured in terms of certain material and energetic effects—the migration and transformation of chemical elements as a function of human activity. However, the actual source of these changes can not be measured in terms of matter or energy. The power of human creativity, is not, itself, measured in material or energy terms, or even in simply biological terms. The changes mankind induces in the material-energetic state of the biosphere or lithosphere exist as a shadow, as a mere expression of a capability, a power, which uniquely lies with the human mind.

It seems that this study opens before us yet another domain of the phenomena of scientific activity, until now exclusively reserved to the speculations of philosophy or religion.

The new form of biogenic migration, at least new to this scale, was provoked, as we see, by the intervention of human reason.

However, it does not dis-

---

29 Emphasis added.

Figure 11

The North American Water and Power Alliance, NAWAPA, would save massive amounts of freshwater from otherwise wastefully running off into the northern Pacific and Arctic oceans, by directing it down through a series of natural trenches, rivers, tunnels, canals, and reservoirs, into the western United States and northern Mexico.

For more see, http://larouchepac.com/nawapaxxi

tinguish itself in any of the other manifestations of biogenic migration, which are connected to other vital functions.

We can, at the same time, establish in a precise way, that human thought changes in a sharp and radical way the course of natural processes, and modifies that which we call the laws of nature.

Consciousness, and thought, despite the efforts of generations of thinkers and wise men, cannot be reduced to either energy or matter, however we define these bases of our scientific thought.

How can consciousness act on the work of natural processes which seem to be entirely reducible to energy and matter?

It is probable that we will not be able to resolve this question until after having radically changed our fundamental physical notions, notions which have undergone and still undergo transformations with a rapidity for which we know of no prior examples in the history of thought.

Thus, the continual demand for progress for mankind takes a fundamentally different form. It is the continual expansion of the creative powers of society, measured in terms of the power of scientific and cultural thought to act upon and transform our planet at higher and higher rates. Society must always move in the direction of higher levels of energy-flux density in terms of physical economics, as measured in the forms of "fire" that can be wielded under the control of scientific thought: the general succession of burning biomass, to coal, petroleum, nuclear fission, thermonuclear fusion, and the prospect of matter-antimatter reactions, is exemplary.

This is coupled with the expansion of human management of more and more of the territory of the Earth. Programs such as the North American Water and Power Alliance (NAWAPA) are exemplary, designed to provide an integrated controlled water management system for much of the North American continent, shifting excess water to where it is needed, and dramatically transforming the biospheric productivity of much of the total land area. Water that is brought inland and participates in plant life is much more likely to evaporate

Figure 12

Mars represents, for mankind, a challenge even more important than Columbus' crossing of the Atlantic Ocean.
Credit: U.S. Geological Survey

and fall back down as rainfall multiple times before returning to the ocean. On average, plants increase this water usage 2.7 times, and more in heavily forested areas.

Despite the great lie of the environmentalist movement—which is an affront to the principle of life itself—continuous, never-ending progress is the only measure by which mankind can justifiably view his actions.

These challenges must be met with the goal of mastering the entire principle of the evolutionary development of the biosphere, and consciously wielding and applying that understanding for the betterment of the Earth itself, and eventually, other planetary bodies as well. Viewed from the perspective of an awaiting barren Mars, such discoveries are crucial, and there is too much progress demanded to waste time with stagnation.

This defines the necessary mission for the progress of mankind, one which would please Vernadsky in celebration of his 150th birthday.

Three Stages of the Phanerozoic Biosphere

	Stage 1: 445 to 250 MYA	Stage 2: 250 to 65 MYA	Stage 3: 65 MYA to Present
Photosynthesis: Ocean	Metabolic Rates / Energy Flux Density (EFD):	Metabolic Rates / EFD:	Metabolic Rates / EFD:
	Technology and Expansion:	**Technology and Expansion:** Cocolythophores and Dinoflagellates emerged as the first forms of eukaryotic phytoplankton to play a dominant role in the oceans.	**Technology and Expansion:** Diatoms developed a urea cycle, increasing more efficient utilization of nitrogen and carbon. They developed a storage vacuole, for storing nutrients.
	Biogenic Migration of Atoms: Cyanobacteria and other prokaryotic phytoplankton are of the green plastid lineage, requiring more iron, zinc, and copper.	**Biogenic Migration of Atoms:** Cocolythophores and dinoflagellates (and diatoms) are of the red plastid lineage, requiring more manganese, cobalt, and cadmium.	**Biogenic Migration of Atoms:** Diatoms made better use of nitrogen, and require silicon, bringing it under a tighter control by ocean life than ever before.
Photosynthesis: Land	**Metabolic Rates / EFD:**	**Metabolic Rates / EFD:**	**Metabolic Rates / EFD:** The energy densities of the fruits of angiosperms are better suited for the higher requirements of mammals, for example. Quick-growing grasses fed the development of grazing mammals.
	Technology and Expansion: Reproduction through spores. Vascular structures to bring water up for vertical growth. Roots to anchor into the ground.	**Technology and Expansion:** The seeds of gymnosperms allows for the penetration into dryer environments, no longer being dependent upon wet environments to reproduce.	**Technology and Expansion:** Angiosperm reproduction makes greater use of other animals as carriers of either fruit or pollen.
	Biogenic Migration of Atoms: ...	**Biogenic Migration of Atoms:** ...	**Biogenic Migration of Atoms:** Grasses require silicon, and have brought the silicon cycle under greater control on land than ever before.
Fungi	**Metabolic Rates / EFD:** Lignin-degrading fungi were rare or absent, leaving biological matter that resists decay for longer in the soils.	**Metabolic Rates / EFD:** The development of fungi with the ability to break down lignin significantly sped up the cycling of carbon and oxygen.	**Metabolic Rates / EFD:** Mushrooms make nutrients accessible to animals, and allow for more specialized spore-production.
Animals: Land	**Metabolic Rates / EFD:** Age of the amphibians.	**Metabolic Rates / EFD:** Age of the reptiles.	**Metabolic Rates / EFD:** Age of the mammals and birds.
	Technology and Expansion of Tetrapods: Moist skin and water-requiring reproductive strategies left amphibians tied to environments near the water.	**Technology and Expansion of Tetrapods:** Dry skin and eggs allowed for the expansion of reptiles into dryer environments.	**Technology and Expansion:** The warmblooded capabilities of birds and mammals allows for their expansion into colder environments.
Animals: Ocean	**Metabolic Rates / EFD:** Stage 1 division of high to low metabolic rates was ~30 / 70. Stage 1 division of predation was ~15 / 85. Shelf bioturbation increased, averaging 2-6 cm, with some regions untouched. "In general terms the Paleozoic dominants were low in individual biomass, their living tissue often arrayed as a thin two-dimensional film coating the skeleton..." (Bambach, 1993)	**Metabolic Rates / EFD:** Stage 2 division of high to low metabolic rates was ~50 / 50. Stage 2 division of predation was ~23 / 77, also associated with Vermij's "Mesozoic marine revolution". Shelf bioturbation increased to the degree that untouched sediments became very rare. "[the] replacement groups in the Mesozoic and Cenozoic and those added into the ecosystem are generally high biomass organisms, often with three dimensional masses of fleshy tissue" (Bambach, 1993)	**Metabolic Rates / EFD:** Stage 3 division of high to low metabolic rates was ~65 / 35. Stage 3 division of predation was ~35 / 65. Shelf bioturbation became so intense that immobile organisms living loosely planted in the sediments could not longer survive . "The energetics of many groups that dominate Cenozoic and modern faunas is greater that that characteristic of Paleozoic dominant groups" (Bambach, 1993)
	Technology and Expansion: Stage 1 percentage of free-swimming species was ~40%.	**Technology and Expansion:** Stage 2 percentage of free-swimming species was ~55%.	**Technology and Expansion:** Stage 3 percentage of free-swimming species was ~80%.

Human Autotrophy

by Vladimir I. Vernadsky

translated from the French by Christine Craig

Translator's Introduction:

Paris held a special place in Vladimir Vernadsky's heart, and he visited it numerous times over the course of his scientific career. His longest stay in Paris was from 1922-1925, the time during which this article was written for the Revue générale des sciences pures et appliquées *(1925) in France.*

It was during this stay that Vernadsky began developing his concept of the noösphere. During this time period he also wrote his book Biosphere, *which he published in Russian in 1926 after returning to Russia.*

In Paris Vernadsky rubbed shoulders, not only with the French (and European) intelligentsia, but with many Russian émigrés who had fled the chaos of the Russian Revolution and its aftermath. The city, with its wealth of scientific and cultural institutions, was incredibly fertile ground for the growth of scientific and social ideas. It was here that the very word noösphere was coined, perhaps by Édouard Le Roy, or maybe Teilhard de Chardin, probably in response to having sat in on Vernadsky's lectures at the Sorbonne. Here, on the Left Bank of the Seine River, the Curies, Pierre and Marie were ensconced at the Radium Institute, while Louis Pasteur lay buried in a vault beneath his Pasteur Institute a few miles away.

Vernadsky had the opportunity to teach for several terms at the Sorbonne (founded in the 13th century), and these lectures may have formed the foundation of his Biosphere. *They no doubt also shaped the ideas he expounds in the present article, where he broaches the subject of the noösphere (a word he does not yet use) in a unique way, by focusing on the idea of human autotrophy: mankind, through scientific advances, freeing himself from reliance on the "ancient material forms of existence," to become "a third branch independent of living matter," along with chemoautotrophs and photoautotrophs.[1]*

I was inspired to translate this work after reading it in the (now-defunct) French magazine Fusion, *Jan.-Feb. 2006. Footnotes, unless indicated as Vernadsky's, are mine.*

---

1. For a fascinating essay on Vernadsky in Paris, please read "Why to Paris," by A. V. Lapo, 2002. URL: http://vernadsky.name/wp-content/uploads/2013/02/Lapo-Pochemu-Parizh-angl.pdf

Sorbonne: Wikipedia user Melusin, Vernadsky: T.B. Pyatibratova, Tambov State Technical University

1 There exists now on the terrestrial surface a great geological force, perhaps cosmic—although planetary action is not generally taken into consideration in concepts of the cosmos, in scientific ideas or those based on science.

This force does not seem to be a new manifestation or special form of energy, nor yet a pure and simple expression of known energy. But it exerts a profound and powerful influence on the course of energetic phenomena on the Earth's surface, and consequently has repercussions, smaller but undeniable, beyond the surface, on the existence of the planet itself.

This force is human reason, the directed and controlled will of social man.

Its manifestation in the environment over the course of myriads of centuries is apparent as one of the expressions of the totality of organisms "of living matter"[2]—of which humanity constitutes but one part.

However over the last several centuries, human society has increasingly distinguished itself from [merely] living matter, by society's action on the surrounding environment. This society becomes in the biosphere, that is to say in the outer envelope of our planet, a unique factor, growing powerfully with great acceleration, a factor which changes—by itself—in a new and rapidly growing manner, the most fundamental mechanisms of the biosphere.

It becomes more and more independent of other forms of life and evolves toward a new vital manifestation.

2 Certainly man seems inseparably tied to living matter—to the entirety of organisms which now exist or which existed before him.

He is linked primarily by his genesis.

No matter how remotely we push into the past, we are sure to find living generations, which are without any doubt genetically linked with each other.

In this past, we can discover with certainty more than ten thousand successive generations, at least—father to son—of *Homo sapiens*, which in their essence cannot be distinguished from us, neither by their character, nor by their exterior, nor by their elevation of thought, nor by the force of their emotions, nor by the intensity of their spiritual life. More than 200 generations have already passed since the era of the birth within human society of the great constructs of religion, science, and philosophy. More hundreds of generations separate us from the times when were laid out the first broad outlines of the works of art, music, myths, magic, which gave birth to religion, to science, and to philosophy.

But the origins of man must be sought even further back in the depths of time. Those ancestors are lost in the mists of the unknown. Their form, their organism, were different than ours; but the essential fact—the succession of generations linked materially, father to son—remains intact. Our connections with these beings so unlike us are concrete. Their past existence is not a fiction.

Francesco Redi (1621-1697)

As far back as our thought or our scientific researches are able to reach into the geological past of the Earth, we encounter the same phenomenon of the existence on the terrestrial crust of one single block of life,[3] uninterrupted, unique. We observe life which is extinguished and renews itself eternally.

About 100 generations have passed since the thinking of the great Greeks focused on this phenomenon, which produced among them the effect of a profound cosmic mystery. It remains for us, their remote descendents, just as it stood before these wise men.

About ten generations before us, the great Florentine naturalist F. Redi—the doctor, poet, man of high morals, a great Catholic Christian—had first expressed a new idea which probably had, from time to time, sprung up in isolated thinkers of past generations, but remained hidden. This revolutionary idea was expressed without, however, coming to the attention of his contemporaries. Their mentality was evidently little prepared. Redi affirmed: All living organisms draw their origins from other living organisms—formally expressed in this form one or two generations after him, by another Italian naturalist, A. Vallisnieri.

This principle of F. Redi was not incorporated into our scientific concepts until the 19th century, almost eight generations after his death. It was a great Frenchman, L. Pasteur, a man of kindred spirit, soulmate of F. Redi, who introduced it definitively into our representation of the cosmos.

Certainly one must represent the genealogy of humankind by the millions of successive generations of beings, which follow, father to son without interruption, and wherein the morphology and functions become modified from time to time. Furthermore, it is extremely likely that life was quite brief for our long-gone ancestors. In measuring the past through the successive generations of man and his ancestors, we arrive at vast numbers surpassing our imagination.

3 Western man has followed the clear path of reason of F. Redi and L. Pasteur, only with reluctance and great effort.

Ideas relating to the eternity of life, to its lack of be-

---

2. On the notion of "living matter" as a group of organisms, see V. Vernadsky: Geochemistry, Paris, Félix Alcan, 1924, p.51. I give in this book a more detailed view of some problems relating to the subject of this article (author's footnote).

---

3. Throughout his article, Vernadsky used the terms *bloc de la vie* and *bloc vivant* (block of life and living block) to indicate the totality of organisms as a group under consideration. This is different from the concept of "life" in itself. I have translated his terms in various ways throughout the translation, while trying to remain faithful to his meaning.

ginning, to the insurmountable difference which exists within the framework of known physical-chemical phenomena, between inanimate matter and living matter, have been in radical discord with his [Western man's] thought-habits, with his worldview. Ideas relating to the beginning and the end of the visible cosmos, of the material universe, as well as to the true unity of all that exists, have profoundly molded his mind.

Oftentimes abiogenesis, that is to say, the genesis of living organisms from inanimate matter without another organism as intermediary, seems logical to the learned; it seems to be a necessary idea for the history of geology and of our planet, and for the scientific explanation of life. They have expressed—with a profound faith—the conviction that the direct synthesis of organisms from scratch out of the material elements will be the inevitable culmination of scientific progress. They don't doubt that there was a moment—if the process follows its course not just in our era—where an organism sprang from the terrestrial crust by a spontaneous change of inanimate matter.

It is necessary to not lose sight of the fact that these conceptions have their root not in the notions of science, but in the domain of religion and philosophy.

Certainly it is *possible* that these conceptions correspond to reality. They cannot yet be considered as refuted by science. But nothing indicates their likelihood. There is nothing to indicate that the problem of abiogenesis is not of the same class as the problems of the quadrature of the circle, the trisection of the angle [by compass and straightedge], perpetual motion, and the philosopher's stone. The inclination of thought to solve these problems has had very important effects. Thanks to it great discoveries have been made—but still the problems are not solvable in the real world.

In order to remain in the domain of science we must declare that:

1. Nowhere have indications of abiogenesis been found in the phenomena which take place or have taken place on the terrestrial crust.

2. Life, such as it presents itself to us in its manifestations and variety, has existed without interruption since the formation of the most ancient geological layers—since the Archean Epoch.

3. Not a single organism exists—among the hundreds of thousands of different species studied—which was not ordered in its genesis exclusively by the principle of F. Redi.

If abiogenesis is not a fiction of the mind, it is only produced outside of known physical-chemical phenomena. Only the discovery of unforeseen phenomena would be able to demonstrate its reality, like the discovery of radioactivity had proved the mass defect in matter and the destruction of the atom, which were only manifested outside of the physical-chemical phenomena studied up until then.

At the present time we are not able to scientifically consider life on our globe otherwise than as an expression of a unique phenomenon which has endured without interruption since the most remote geological times whose clues we have been able to study. Living matter has endured throughout all this time separated from inanimate matter. Man is irrevocably linked to the same totality of life with all the living beings which exist or which have existed.

4 Man is also linked to this totality by his nutrition. This new connection, as intimate and as indispensable as it is, is not of the same order as the uninterrupted succession of the generations of living beings. This connection doesn't appear to us as a profound natural process, immutable, indispensable to life like that which is expressed by the Redi principle.

It is true that this connection is part of a great geochemical phenomenon—of the circulation of the chemical elements in the biosphere because of the nutrition of organized beings. This connection has perhaps changed, yet without affecting the stabilty of the totality of life. In the paleontological history of the biosphere, there are serious indications of an analogous shift which had already taken place in the course of time, in the evolution of certain groups of bacteria—invisible and minute beings, but with strong geochemical power.

Man's dependence on the living for his nutrition presently rules all of his existence. A change in regime, were it to come, would have immense consequences. The crucial fact, at the present moment, is the potential which is proper for man to preserve his existence, to construct and keep intact his unique body through the assimilation, either of other organisms, or of products of their life. The chemical compounds thus formed in the terrestrial crust are necessary and indispensable for existence, but the human organism does not have the means to produce them himself. He must look for them in his living environment, annihilating other living beings or exploiting their biochemical work. He dies if he finds himself upon the terrestrial surface in the absence of other living beings, which constitute his nourishment.

It is clear that all human life, all societies developed in the course of history, are controlled by this necessity. In the last analysis, it is this irresistible need which governs the human world, which shapes all of its history and all of its existence.

It is famine, in the end, which is the pitiless factor, the terrible agent of the social edifice. Social equilibrium is only achieved by incessant labor, and it is always unstable. The great disruptions of society, the crimes perpetrated on this terrain always have disastrous consequences.

Our civilization in this respect finds itself always at the brink of a precipice. At present hundreds of thousands of men die or languish in Russia because of lack of nourishment and millions of others—more than 10-15 millions—have been victims of social wrongs. Never has the precariousness of human existence been so clear and the specter of disgrace and decadence so alive in the spirit of disorder.

5 Only recently—less than five generations separate us from those times—has man begun to understand the intricate and very special structure of the living system in which he appears.

And as yet the consequences of this structure—enormous social and political consequences—have not penetrated his thought.

One can see this plainly in considering current social ideas that are promulgated around us and which set the world into motion. These ideas reside fundamentally outside of today's science. They are the expression of the past in the exact sciences, corresponding to the science of one hundred years ago! All the progress in science of the 19th- and 20th centuries have had but a feeble influence on contemporary social thought. The exact sciences have been transformed from the bottom up and their antagonism with social ideas has become greater and greater. Not just the masses—but those who lead and inspire as well—belong in their thinking and their scientific baggage to a long-past stage of scientific evolution. Humanity, in its actual social development is in large part governed by ideas which conform little to reality and express the scientific thinking and knowledge of vanished generations of the past.

A profound change of social and political ideas, because of fundamental new acquisitions in natural science, in the exact sciences, is imminent, and it is already making an appearance. The problems of nutrition and of production must be reexamined. This change will necessarily be followed by an upheaval in the very social principles which direct opinion. The slow infiltration of scientific acquisitions into life and into thought is a habitual and general trait in the history of science.

6 The new foundations of our present representation of nutrition were achieved in the years before the end of the 18th century by the efforts of a small elite of humanity who transformed our conception of the world without having been understood or valued by their contemporaries.

They were, first, Lord H. Cavendish of London, the richest man in his country, misanthrope and ascetic of science; A. L. Lavoisier, financier and experimentalist, a profound and lucid thinker, whose assassination is an indelible shame for humanity; the ardent theologian and radical Englishman J. Priestly, persecuted and misunderstood, who by luck escaped death when the mob burned and destroyed his house, his laboratory, and his manuscripts, and who had to flee his country; the Genevese aristocrat, representative of a family where high scientific culture was hereditary, Th. de Saussure; the profound Dutch naturalist and doctor J. Ingen-Housz who, because he was Catholic, could

Wilhelm Pfeffer, German plant physiologist (1845-1920)

not make a career in his country and worked in Vienna and England.... They were followed by many researchers in all countries.

One or two generations after these pioneers—around 1840—their thinking had definitely penetrated science and was expressed lucidly and fully in Paris by J. Boussaingault and J. Dumas, and at Giessen in Germany by J. Liebig.

A major effect of immense impetus was unleashed by this labor.

7 The living system—the world of organisms—seems *double* in function and position in the crust.

The greater part of living matter, the world of green plants, depending only on inanimate matter, is independent of other organisms. The green plants are able to create for themselves the necessary substances for their life in utilizing the inorganic chemicals in the crust. They take the gasses and aqueous solutions from the surrounding environment and construct for themselves innumerable carbon and nitrogen compounds—hundreds of thousands of different substances—which are incorporated into the composition of their tissues.

German physiologist W.[4] Pfeffer distinguished organisms which possess these abilities by the name of *autotrophic organisms*, because they were only dependent on themselves for their nutrition. He named *heterotrophic*

---

4. Vernadsky mistakenly had Pfeffer's first initial as J., but is clearly referring to the great German pioneer in plant physiology, Wilhelm Pfeffer.

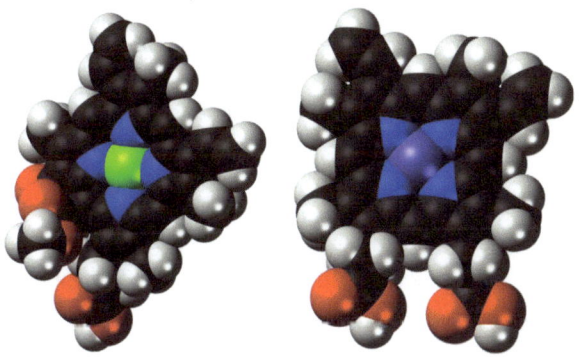

Three-dimensional space-filling images of the porphyrin molecule common to both chlorophyll and hemoglobin. On the left is the chlorophyll-a porphyrin molecule with its magnesium center. On the right is the heme porphyrin molecule with its iron center.

those organisms which depended, for their nutrition, on other organisms, utilizing their chemical products. They are able only to change chemical compounds made outside themselves, which they appropriate for their life, but cannot construct for themselves.

There exist green organisms whose nutrition is mixed, organisms which in part make the necessary chemical compounds, and use the substances of inanimate matter, and in part obtain it—as with parasites—by exploiting other organisms. These beings, numerous in living nature, are the *mixotrophs* of Pfeffer. Mistletoe is a well-known example.

In the final analysis autotrophic green organisms—green plants—form the foundation for the living system. The world as diverse as the mushrooms, the millions of animal species, humankind—cannot exist except as a consequence of their biochemical work. This work would not be possible except by the grace of the innate property of these organisms to transform the energy of solar radiation to chemical free energy.

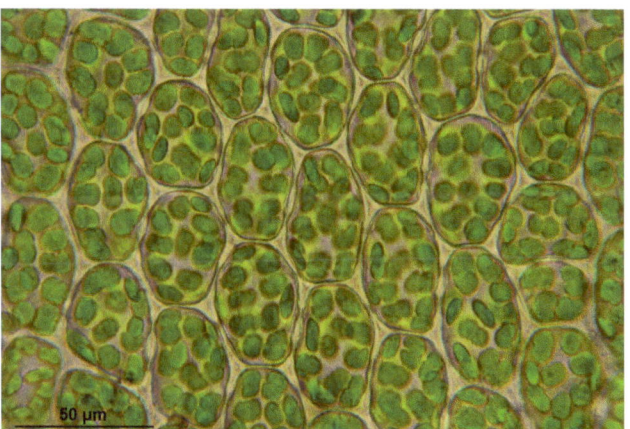

50 μm

wikimedia: Hermann Schachner

Cells of the moss Plagiomnium affine *showing numerous plastids per* cell.

It is clear that life is not a simple terrestrial phenomenon, but manifests itself as a cosmic phenomenon in the history of our planet, in so far as the principle of Redi corresponds to reality.

And furthermore it follows that the living system is not an assemblage of isolated individuals, an assemblage owing to chance, but exhibits a mechanism where the constituents have functions which influence and coordinate it.

8 Autotrophic green matter is able to perform its proper function in this mechanism thanks to its elaboration of a green substance with very specific and remarkable properties—*chlorophyll*. It is a complex compound which contains atoms of magnesium and has a molecular structure, containing carbon, oxygen, hydrogen, and nitrogen, that is quite similar to that of the red hemoglobin in our blood, where the magnesium is replaced by iron.

Chlorophyll, whose structure and chemical properties are beginning to become clear, is produced in plants within special tiny microscopic granules—the plastids—dispersed throughout the cells. These plastids only originate from the division of other plastids. The organism is unable to obtain them otherwise. This demonstrates a remarkable fact, which indicates a general phenomenon analogous to that expressed by F. Redi's principle. No matter how far we push back into the past—we see the formation of chlorophyllic plastids brought about exclusively by previously formed plastids.

Thanks to plastids of chlorophyll, the organism of green plants is able to pass down its life to other organisms.

If we only considered their nutrition—green plants would be able to exist alone on the surface of our planet.

9 The repercussion of the existence of autotrophic organisms with chlorophyllic function on the surface of the Earth is immense.

Not only is it they which give birth to all other organisms and humankind—but they regulate the chemistry of the terrestrial crust. One can get an idea of the magnitude of this phenomenon by recalling some numerical facts.

The verdure of our gardens, our fields, forests, and prairies surround us. Seen from another planet, from cosmic space, Earth would have a green tint. But that mass of chlorophyll represents but a part. The greatest portion of chlorophyll is invisible to us. It lies in the uppermost layers of the worldwide ocean at depths of up to approximately 400 meters. It is contained in innumerable myriads of unicellular, invisible algae each of which gives birth in the course of two or three daily rotations of our planet, to a new generation, which begins to reproduce itself. In this way, if they did not figure into the nutrition of other beings, their number would become prodigious and fill the worldwide ocean.[5]

The existence of free oxygen in our atmosphere and in the waters is the expression of the chlorophyllic function. All the free oxygen of the globe is a product of green plants. If green plants no longer existed, in a few hundreds of years there would not remain a trace of free oxygen on the surface of the Earth, and in the end chemical transformations would capture it all.[6]

The mass of free oxygen of the surface of the Earth corresponds to 1.5 quadrillion (1×10^{15}) metric tons. That number gives only an idea of the geochemical importance of life.

The amount of chlorophyll produced in green plants necessary to keep free oxygen at this level corresponds to many billions of tons at least, existing at each moment in the bodies of autotrophic plants.

S. N. Winogradsky, the Ukrainian-Russion microbiologist and soil scientist (1856-1953).

10 It has been more than thirty years since the Russian biologist S. N. Winogradsky introduced into this situation a new and important attribute which demonstrates the already-great complexity of the living system.

He discovered the existence of autotrophic living beings without chlorophyll. These are invisible beings, bacteria which teem in the soils, in the superficial parts of the crust, and penetrate the floor of the worldwide ocean.

Notwithstanding their smallness, thanks to their prodigious reproduction, their importance in the economy of nature is huge. This enormous reproduction—comparable with that of the unicellular green algae—obliges us to consider their existence as a phenomenon on the order of that of green plants.

Certainly the number of species of autotrophic bacteria is small, not more than a hundred, while that of green plants is close to 180,000. But whereas in a day each bacterium is able to engender many trillions of individuals, one green unicellular alga, which of all the green plants reproduces the most rapidly, cannot produce in the same interval of time but a few, and generally much less—say one sole individual in two or three days.[7]

The bacteria discovered by S. Winogradsky are independent in their nutrition not only of other organisms, but of solar radiation. In the construction of their bodies they use chemical energy from terrestrial chemical compounds—the minerals—rich in oxygen.

They produce by means of this decomposition—and by virtue of the syntheses which are their consequence—an immense geochemical work. Their role is very great in the history of carbon, sulfur, nitrogen, iron, manganese, and probably many other elements of our globe.

It is certain that they belong to the same life group as the other organisms, because they get their nutrition from these last and use their waste. We are led to think that the connection is very close, that they belong in this genetic group.

One can consider them as very specialized derivatives of green plants, as is done for non-chlorophyllic plants in general, yet without excluding the possibility of seeing in them representatives of the ancestors of chlorophyll-producing beings.

In our present state of knowledge, the first hypothesis seems most likely. Nevertheless, one must always take into account that these organisms of S. Winogradsky play a preponderant role in phenomena of the superficial modification of terrestrial minerals. These modifications seem to be immutable over the course of the geological history of our planet. They have not changed since the Archaic Era.

11 Man is a heterotrophic social animal. He can only exist in the presence of other organisms, especially green plants.

His existence on our planet is clearly distinguished all the same, from that of all the other organized beings. Reason, which distinguishes man within the assemblage of living matter, gives living matter remarkable characteristics, profoundly changing its [living matter's] action on the environment.

The genesis of man was a singular event, unique in geological history, which had no analog in the preceding myriads of centuries.

From the scientific standpoint, one must consider it as the consequence of a long natural process, of which the beginning is lost to us, but which has lasted without interruption over the course of all of geologic time. Until now, no scientific theory has been able to encompass the paleontological evolution of organized beings, of which the latest important expression has been the genesis of man.

---

5. Most, but certainly not all of these "invisible algae" would now be placed among the prokaryotes as cyanobacteria.

6. This was written in 1925, and shows a clear understanding of the primary role of photoautotrophic organisms in the generation of the present atmosphere of Earth, which he addresses again in his *Biosphere*, published a year later.

7. If the chemoautotrophic bacterium divided once every half-hour, in 24 hours it would produce about 281.5 trillion individuals. If the eukaryotic unicellular alga divided once every 24 hours it would produce two individuals.

Left: Lori Johnston, RMS Titanic Expedition 2003, NOAA-OE, Right: Courtesy of NOAA/Institute for Exploration/University of Rhode Island (NOAA/IFE/URI)

"Rusticles" feasting on the largesse of the noösphere. These are consortia of bacteria and fungi enjoying a 100-year feast on the iron parts of the sunken Titanic 3.8 km beneath the ocean surface. Chemoautotrophs present in this living community are capable of deriving energy from oxidizing the iron deposited on the deep, dark, and oxygen-poor terrain of the ocean bottom after the RMS Titanic sank.

We are unable to represent the genetic change of the living system—the extinction and generation of innumerable species—except under an empirical generalization—that of the evolution of species.

For a man of science, the empirical generalization is the foundation of all knowledge, its form the most certain. But, to connect it to other facts and empirical generalizations, the learned man must avail himself of theories, axioms, models, hypotheses, abstractions. We have but an imperfect sketch in this domain.

It is clear that there exists a determined direction in the paleontological evolution of organized beings, and that the appearance of understanding, of reason, of coordinated will on the terrestrial surface—this manifestation of man—cannot be a game of chance. But it is impossible for us at present to give an explanation of this phenomenon, that is to say of the logical connection with our abstract scientific construct of the world—based on these models and these axioms.

12 Man is profoundly distinguished from the other organisms by his action on the environment. This distinction, which was great from the beginning, has become immense with the passage of time.

The action of other organisms is almost exclusively determined by their nutrition and their growth and increase. The sole fact of the formation of free oxygen is sufficient to appreciate the planetary importance of their nutrition. And it is one fact among thousands of others. The formation of coal, petroleum, iron-bearing minerals, humus, calcites, coral islands, are isolated cases—among thousands of others—of the manifestation of their increase.

Mankind certainly acts in the same way as all of these organisms. But his mass is completely negligible in comparison with the totality of living matter and the direct manifestations in living nature of his nutrition and his increase are almost nothing. The wise Austrian economist L. Brentano has given a very clear representation of the scale of humanity within the environment. If one assigns to each human individual a square meter, and if one brings together all the humans existing on the terrestrial surface—the surface that they would occupy would not exceed that of Lake Constance.[8]

It is clear that the manifestation of such a living mass considered on the scale of geological phenomena would be negligible.

Reason changes all. Through it, man utilizes material in the environment—inanimate or living—not only for the building of his body, but also for social life. And this usage has become a great geological force.

Thought, by its existence, introduces into the crustal mechanisms a powerful process having no analog before the appearance of man.

13 Man is the *Homo faber* of M. H. Bergson. He changes the aspect, the chemical and mineralogical composition of his living environment. His living environment is the whole of the surface of Earth.

His action becomes stronger and more coordinated with each passing century. The naturalist must acknowledge a natural process of the same order as all of the other geological manifestations. This process is necessarily regulated by the principle of inertia—it will follow its

---

8. Lake Constance, 571 km² in size, lies between Germany (Bavaria) and Switzerland at the foot of the Alps, and is fed by the Rhine River.

A patchwork of farmland in northwest Minnesota along the Buffalo River

course regardless, if forces don't exist which oppose it or which take it to a potential state.

The discovery of agriculture, made over 600 generations before us, decided the path of humanity. By controlling the life of the autotrophic green organisms on the terrestrial surface man gained leverage, with immense consequences for the history of the planet. Man has become by this fact master of all living matter, and not just green plants, since the existence of all beings is controlled by the green plants. Little by little he changed living matter by the decisions—the goals—of his reason.

Through agriculture, he was liberated—in his nutrition—from the natural living environment, of which all the other organized beings are naught in this respect but impotent processes.

14 Relying on this great conquest, man has annihilated "virgin nature." He has introduced immense quantities of new, unknown chemical compounds and new forms of life—races of animal and plants.

He has changed the course of all of geochemical reactions. The face of the planet became new and found itself in a state of continual upheaval.

But man has not yet succeeded in gaining, in this new environment, the security necessary for his life.

In his social organization, existence itself, for the majority is precarious, the distribution of wealth does not give to the great mass of humanity the means of a life conforming to moral and religious ideals.

New, troubling events, which relate to the bases of his existence, are let loose in these recent times.

The reserves of natural resources decrease visibly. If their usage grows with the same force, the situation will become grave. In two generations one would detect a scarcity of iron; petroleum would also quickly become scarce; in a few generations, the question of coal would become tragic. It is the same for most of the other natural resources. The dearth of coal would be particularly grave, because it is coal which procures for man the energy necessary for society in its present form.

This is an inevitable phenomenon, because man uses the stores of natural resources which were formed throughout myriads of centuries and which could not be replenished except in the same length of time. These reserves are necessarily restricted. Similarly if one found other unknown sources, or if one used the less rich or deeper concentrations—one would only push back the date of the critical period—but the troubling problems would remain unresolved.

For generations, profound thinkers have perceived the necessity of radical social means, of scientific acquisitions of a new order to rein in the imminent danger. At the beginning of the last century, the imminent scarcity of natural resources was not yet perceived, because the energy at man's disposal in this era was still largely connected to ancient material forms of existence—to the life and works of men, of plants, and of animals. Nevertheless already the founders of socialism—particularly Count H. de Saint-Simon, W. Godwin, and R. Owen—understood the primary importance of science, the impossibility of resolving the social question while using only the resources which existed in their day, without augmenting, by science, the means of human power.

It was truly a scientific socialism in a sense which has since been forgotten.

The problem which is posed at this moment before humanity clearly goes beyond the social ideology, which has since been elaborated by the socialists and communists of all schools, who in their constructs have allowed the vivifying spirit of science—its social role—to elude them. Our generation has been victim of an application of this ideology in the course of tragic events in my country—one of the richest in natural resources—of which the results were death and famine for the multitude and economic failure of the communist system which seems undeniable. But the failure of socialism seems more profound. It presents in general the social problem from a too-restricted viewpoint, which does not correspond with reality; it remains superficial.

15 To resolve the social question it is necessary to plumb the foundations of human power—to change the form of nourishment and the sources of energy which man uses.

Precisely on these two points, little by little the thoughts of researchers are engaged. Here one is on solid ground. Not only can there be no doubt of the possibility of solving these two problems, but it is also clear that they will inevitably be solved in a very short time, even in comparison with the human lifespan.

The solution to these problems is taking shape as a result of scientific progress outside of all social preoccupation. After generations, science, in its quest for truth, is forced to discover new forms of energy in the world and great organic chemical syntheses. It labors with very insufficient means, the only ones available in human society today, where the situation is in striking contradiction with its [science's] real role as producer of wealth and of human power.

This scientific movement can be accelerated by creating new methods of research; it can't be stopped. Because there is not a force in the world which can shackle human understanding in its march, once it has understood, as in the present case, the scope of the truths which are opened before it.

16 Until now, the power of fire in its multiple forms was almost the sole source of energy for society. Man obtained it by the combustion of other organisms or their fossil remains.

Some decades ago, he began systematically to replace it by other sources of energy, independent of life—first by hydropower. The quantity of hydropower—the motive force of water—existing on the terrestrial surface was measured. And it was seen that, large as it seems, it is not sufficient by itself for societal requirements.

But the reserves of energy which are at the disposal of reason are inexhaustible. The force of the tides and ocean waves, radioactive atomic energy, solar heat are able to give us all the power needed.

The introduction of these forms of energy into life is a matter of time. It depends on problems whose solutions present nothing impossible.

The energy thus obtained will not have practical limits.

In directly utilizing the energy of the sun, man is made master of the source of energy of the green plant, of the form that he now uses through the intermediary of the latter in his nourishment and as fuel.

17 The synthesis of foodstuffs, freed from the intermediary of organized beings, when accomplished, will change human prospects.

It grips the imagination of the learned after the great successes of organic chemistry; in fact it presents a hidden but always vibrant aspiration of laboratories. It is never lost from view. If the great chemists only express it from time to time, like the able M. Berthelot, it is because they know that the problem will not be resolved before the undertaking of a long preliminary work. The work is carried out systematically, but must be the labor of many generations, considering the great poverty of science within our social structure.

One generation has already disappeared since the death of M. Berthelot. We are much closer to this supreme goal than we were during his lifetime. We can follow its slow but incessant progress. After the brilliant work of the German chemist E. Fischer and his school on the structure of albumin and of the carbohydrates, there can be no doubt of its eventual success.

During the Great War, the problem was often envisaged in various countries in its practical aspect and

Deviantart user Shefu-de-combinat

Ammonia processing by the Haber-Bosch process was one of the most significant steps toward human autotrophy in modern history. It can be argued that the synthesis of ammonia from atmospheric nitrogen and hydrogen without the required intervention of microorganisms, can be credited with enabling half of our 7 billion earthlings to be alive today. Furthermore, half the nitrogen present in the average human body today, came from Haber-Bosch ammonia, not from life-derived nitrogen-fixing processes. The green revolutions were fueled by Haber-Bosch nitrogen fixation, which is today fueled mainly by natural gas, but could be integrated into nuplexes using fourth-generation nuclear technology to boost this technology to the next level.

thewhici faith in its imminent solution took deep root among the learned.

Certainly it often happens that a scientific discovery is lost or doesn't find its practical application, its introduction into life, until long after it was first made. But we can be confident that the synthesis of food will not meet such a fate.

We await the discovery of this synthesis, and its great consequences to life will immediately be manifested.

18 What would be the significance of the synthetic production of nutriments to human life and to the life of the biosphere?

By its accomplishment man would free himself from living matter. From a social heterotrophic being, he would become an *autotroph*.

The repercussion of this phenomenon within the biosphere would be immense. It would signify the schism of the block of life, the creation of a third branch independent of living matter. By this feat there would appear on the terrestrial surface, and for the first time in the geological history of the globe, an *autotrophic animal*.

Today, it is difficult, perhaps impossible, for us to grasp the geological consequences of this event—but it is clear that it would be the culmination of a long paleontological evolution, which would represent, not an action of the free will of humanity, but the manifestation of a natural process.

By this achievement, human understanding would produce not only a great social effect, but a great geological phenomenon.

19 The repercussion of this synthesis in human society shall certainly touch us with ever-greater force. Will it bring good or will it bring new desolations to the human species? We don't know. But the course of phenomena—the future—will be perhaps controlled by our will and by our reason. We must prepare to understand the consequences of the actions of this inevitable discovery.

Only isolated thinkers sense the approach of this new age. They see these consequences differently.

One finds the expression of these intuitions in works of fiction. The future seems troubled and tragic for some (*Histoire de quatre ans*, by D. Halévy), at the same time that others see it as great and beautiful (*Auf zwei planeten*, from the profound German thinker and historian of ideas, K. Lasswitz).

The naturalist can only contemplate this discovery with a great tranquility.

He sees in its accomplishment the outcome of a grand natural process which has endured for millions of years and which gives no sign of dissipating. It is a creative process, and not anarchic.

Indeed, man's path is always formed in great part by man himself. The creation of a new autotrophic being will give him possibilities which have been lacking for the accomplishment of his secular moral aspirations; it will open for him the path to a better life.

V. Vernadsky
Member of the Russian Academy of Sciences
General Review of Pure and Applied Sciences, 1925.

Some Words About The Noösphere[1]

by Vladimir I. Vernadsky

The following article was written in December 1943. An abridged version was published in English in the American Scientist, *January 1945, translated by the author's son, Dr. George Vernadsky of Yale University. The full translation (including portions of George Vernadsky's translation) is provided here by Rachel Douglas of* Executive Intelligence Review, *translated from the Russian edition contained in Vernadsky's book* Biosfera *(Moscow: Mysl Publishing House, 1967).*

Subheads have been added.

Vladimir Ivanovich Vernadsky (1863-1945), who developed the concept of the biosphere and how man's creativity has changed it into the noösphere.

We are approaching the climax in the Second World War. In Europe war was resumed in 1939 after an intermission of twenty-one years; it has lasted five years in Western Europe, and is in its third year in our parts, in Eastern Europe. As for the Far East, the war was resumed there, much earlier, in 1931, and is already in its 12th year. A war of such power, duration, and strength is a phenomenon unparalleled in the history of mankind and of the biosphere at large. Moreover, it was preceded by the First World War which, although of lesser power, has a causal connection with the present war.

In our country that First World War resulted in a new, historically unprecedented, form of statehood, not only in the realm of economics, but likewise in that of the aspirations of nationalities. From the point of view of the naturalist (and, I think, likewise from that of the historian), an historical phenomenon of such power may and should be examined as a part of a single great terrestrial *geological* process, and not merely as a *historical* process.

In my own scientific work, the First World War was reflected in a most decisive way. It radically changed my *geological conception of the world.* It is in the atmosphere of that war that I have approached a conception of nature, at that time forgotten and thus new for myself and for others, a geochemical and biogeochemical conception embracing both nonliving and living nature from the same point of view.[2] I spent the years of the First World War in my uninterrupted scientific creative work, which I have so far continued steadily in the same direction.

Twenty-eight years ago, in 1915, a "Commission for the Study of the Productive Forces" of our country, the so-called KEPS, was formed at the Academy of Sciences. That commis-

sion, of which I was elected president, played a noticeable role in the critical period of the First World War. Entirely unexpectedly, in the midst of the war, it became clear to the Academy of Sciences that in Tsarist Russia there were no precise data concerning the now so-called strategic raw materials, and we had to collect and digest dispersed data rapidly to make up for the lacunae in our knowledge.[3] Unfortunately by the time of the beginning of the Second World War, only the most bureaucratic part of that commission, the so-called Council of the Productive Forces, was preserved, and it became necessary to restore its other parts in a hurry.

By approaching the study of geological phenomena from a geochemical and biogeochemical point of view, we may comprehend the whole of the circumambient nature in the same atomic aspect. Unconsciously, such an approach coincides for me with what characterizes the science of the 20th Century and distinguishes it from that of past centuries. *The 20th Century is the century of scientific atomism.*

At that time, in 1917-1918, I happened to be, entirely by chance, in the Ukraine,[4] and was unable to return to Petrograd until 1921. During all those years, wherever I resided, my thoughts were directed toward the geochemical and biogeochemical manifestations in the circumambient nature, the biosphere. While observing them, I simultaneously directed both my reading and my reflection toward this subject in an intensive and systematic way. I expounded the conclusions arrived at gradually, as they were formed, through lectures and reports delivered in whatever city I happened to stay, in Yalta, Poltava, Kiev, Simferopol, Novorossiysk, Rostov, and so on. Besides, in almost every city I stayed, I used to read everything available in regard to the problem in its broadest sense. I left aside, as much as I could, all philosophical aspirations and tried to rest only on firmly established scientific and empiric facts and generalizations, occasionally allowing myself to resort to working scientific hypotheses.

Instead of the concept of "life," I introduced that of "living matter," which now seems to be firmly established in science. "Living matter" is the totality of living organisms. It is but a scientific empirical generalization of empirically indisputable facts known to all, observable easily and with precision. The concept of "life" always steps outside the boundaries of the concept of "living matter"; it enters the realm of philosophy, folklore, religion, and the arts. All that is left outside the notion of "living matter."

In the thick of life today, intense and complex as it is, a person practically forgets that he, and all of mankind, from which he is inseparable, are inseparably connected with the biosphere—with that specific part of the planet, where they live. It is customary to talk about man as an individual who moves freely about our planet, and freely constructs his own history. Hitherto, neither historians, scientists in the humanities, nor, to a certain extent, even biologists, have consciously taken into account the laws of the nature of the biosphere—the envelope of Earth, which is the only place where life can exist. Man is elementally indivisible from the biosphere. And this inseparability is only now beginning to become precisely clear to us. In reality, no living organism exists in a free state on Earth. All of these organisms are inseparably and continuously connected—first and foremost by feeding and breathing—with their material-energetic environment.

The outstanding Petersburg academician Caspar Wolf (1733-1794), who dedicated his whole life to Russia, expressed this brilliantly in his book, published in German in St. Petersburg in 1789, the year of the French Revolution: *On the Peculiar and Efficient Force, Characteristic of Plant and Animal Substance.* Unlike the majority of biologists of his day, he relied upon Newton, rather than Descartes.[5]

Mankind, as living matter, is inseparably connected with the material-energetic processes of a specific geological envelope of the Earth—its *biosphere.* Mankind cannot be physically independent of the biosphere for a single minute.

The 'Huygens Principle'

The concept of the "biosphere," i.e., "the domain of life," was introduced in biology by Lamarck (1744-1829) in Paris at the beginning of the 19th Century, and in geology by Edward Suess (1831-1914) in Vienna, at the end of that century.[6] In our century there is an absolutely new understanding of the biosphere. It is emerging as a *planetary* phenomenon that is *cosmic in nature.* In biogeochemistry we have to consider that life (living organisms) really exists not on our planet alone, not only in the Earth's biosphere. It seems to me that this has been established beyond a doubt, so far, for all the so-called terrestrial planets, i.e., for Venus, Earth, and Mars.[7] At the Biogeochemical Laboratory of the Academy of Sciences in Moscow, which has been renamed the Geochemical Problems Laboratory, in collaboration with the Microbiology Institute of the Academy of Sciences (director—Corresponding Academician B.L. Isachenko), we identified *cosmic* life as a matter for current scientific study already in 1940. This work was halted because of the war, and will be resumed at the earliest opportunity.

The idea of life as a cosmic phenomenon has been found in the scientific archives, including our own, for a long time. Centuries ago, in the late 17th Century, the Dutch scientist Christiaan Huygens (1629-1695), in his last work, *Cosmotheoros,* which was published posthumously, formulated this scientific question. The book was published in Russian twice in the first quarter of the 18th Century, on the initiative of Peter I.[8] In this book, Huygens established the scientific generalization that "life is a cosmic phenomenon, in some way sharply distinct from nonliving matter." I recently named this generalization "the Huygens principle."[9]

By weight, living matter comprises a minute part of the planet. This has evidently been the case throughout all geological time, i.e., it is geologically eternal.[10] Living matter is concentrated in a thin, more or less continuous layer in the troposphere on dry land—in fields and forests—and permeates the entire ocean. In quantity, it measures no greater than tenths of a percent of the biosphere by weight, on the order of close to 0.25 percent. On dry land, its continuous mass reaches to a depth of probably less than 3 kilometers on average. It does not exist outside the biosphere.

In the course of geological time, living matter morphologically changes, according to the laws of nature. The history of living matter expresses itself as a slow modification of the forms of living organisms, which genetically are uninterruptedly connected among themselves from generation to generation. This idea had been rising in scientific research through the ages, until, in 1859, it received a solid foundation in the great achievements of Charles Darwin (1809-1882) and [Alfred R.] Wallace (1822-1913). It was cast in the doctrine of the evolution of species of plants and animals, including man. The evolutionary process is a characteristic only of living matter. There are no manifestations of it in the nonliving matter of

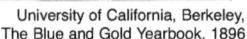
University of California, Berkeley,
The Blue and Gold Yearbook, 1896

U.S. Geological Survey

The American geologist Joseph Le Conte (1823-1901), at left, developed the idea that living matter was evolving in a definite direction, which he called the Psychozoic era. James Dwight Dana (1813-1895), a geologist, mineralogist, and biologist, developed a similar idea, which he called cephalization. Dana was a member of the Wilkes Expedition.

our planet. In the Cryptozoic era, the same minerals and rocks were being formed which are being formed now.[11] The only exceptions are the bio-inert natural bodies connected in one way or another with living matter.[12]

The change in the morphological structure of living matter, observed in the process of evolution, unavoidably leads to a change in its chemical composition. This question now requires experimental verification. In collaboration with the Paleontology Institute of the Academy of Sciences, we included this problem in our planned work in 1944.

While the quantity of living matter is negligible in relation to the nonliving and bio-inert mass of the biosphere, the biogenic rocks constitute a large part of its mass, and go far beyond the boundaries of the biosphere. Subject to the phenomena of metamorphism, they are converted, losing all traces of life, into the granitic envelope, and are no longer part of the biosphere. The granitic envelope of the Earth is the area of former biospheres.[13] In Lamarck's book, *Hydrogeologie* (1802), containing many remarkable ideas, living matter, as I understand it, was revealed as the creator of the main rocks of our planet. Lamarck never accepted Lavoisier's (1743-1794) discovery. But that other great chemist, J.B. Dumas (1800-1884), Lamarck's younger contemporary, who did accept Lavoisier's discovery, and who intensively studied the chemistry of living matter, likewise adhered for a long time to the notion of the quantitative importance of living matter in the structure of the rocks of the biosphere.

Cephalization—the Arrow of Evolution

The younger contemporaries of Darwin, J[ames] D[wight] Dana (1813-1895) and J[oseph] Le Conte (1823-1901), both great American geologists (and Dana, a mineralogist and

biologist as well) expounded, even prior to 1859, the empirical generalization that *the evolution of living matter is proceeding in a definite direction.* This phenomenon was called by Dana "cephalization," and by Le Conte the "Psychozoic era." Dana, like Darwin, adopted this idea at the time of his journey around the world, which he started in 1838, two years after Darwin's return to London, and which lasted until 1842.[14]

It should be noted here that the expedition during which Dana reached his conclusions about cephalization, coral reefs, and so on, was historically associated with the researches on the Pacific Ocean, done on ocean voyages by Russian sailors, notably Kruzenshtern (1770-1846). Published in German, they inspired the American lawyer John Reynolds to organize the first such American scientific sea voyage.[15] He began to work towards this in 1827, when an account of Kruzenshtern's expedition came out in German. Only in 1838, 11 years later, did his persistent efforts result in this expedition taking place. This was the Wilkes expedition, which conclusively proved the existence of Antarctica.

Empiric notions of a definite direction of the evolutionary process, without, however, any attempt theoretically to ground them, go deeper into the 18th Century. Buffon (1707-1788) spoke of the "realm of man," because of the geological importance of man. The idea of evolution was alien to him. It was likewise alien to Agassiz (1807-1873), who introduced the idea of the glacial period into science. Agassiz lived in a period of an impetuous blossoming of geology. He admitted that, geologically, the realm of man had come, but, because of his theological tenets, opposed the theory of evolution. Le Conte points out that Dana, formerly having a point of view close to that of Agassiz, in the last years of his life accepted the idea of evolution in its then-usual Darwinian interpretation.[16] The difference between Le Conte's "Psychozoic era" and Dana's "cephalization" thus disappeared. It is to be regretted that, especially in our country, this important empirical generalization still remains outside the horizon of our biologists.

The soundness of Dana's principle, which happens to be outside the horizon of our paleontologists, may easily be verified by anyone willing to do so on the basis of any modern treatise on paleontology. The principle not only embraces the whole animal kingdom, but likewise reveals itself clearly in individual types of animals. Dana pointed out that in the course of geological time, at least 2 billion years and probably much more, there occurs an irregular process of growth and perfection of the central nervous system, beginning with the crustacea (whose study Dana used to establish his principle), the mollusca (cephalopoda), and ending with man. It is this phenomenon he called cephalization. The brain, which has once achieved a certain level in the process of evolution, is not subject to retrogression, but only can progress further.

The Noösphere Comes of Age

Proceeding from the notion of the geological role of man, the geologist A.P. Pavlov (1854-1929) in the last years of his life used to speak of the *anthropogenic era,* in which we now live. While he did not take into the account the possibility of the destruction of spiritual and material values we now witness in the barbaric invasion of the Germans and their allies, slightly more than 10 years after his death, he rightfully emphasized that man, under our very eyes, is becoming a mighty and ever-growing geological force. This geological force was formed quite imperceptibly over a long period of time. A change in man's position on our planet (his material position first of all) coincided with it. In the 20th Century, man, for the first time in the history of the Earth, knew and embraced the whole biosphere, completed the geographic map of the planet Earth, and colonized its whole surface. *Mankind became a single totality in the life of the Earth.* There is no spot on Earth where man can not live if he so desires. Our people's sojourn on the floating ice of the North Pole in 1937-1938 has proved this clearly. At the same time, owing to the mighty techniques and successes of scientific thought, radio and television, man is able to speak instantly to anyone he wishes at any point on our planet. Transportation by air has reached a speed of several hundred kilometers per hour, and has not reached its maximum. All this is the result of "cephalization," the growth of man's brain and the work directed by his brain.

The economist, L. Brentano, illuminated the planetary significance of this phenomenon with the following striking computation: If a square meter were assigned to each man, and if all men were put close to one another, they would not occupy the area of even the small Lake of Constance between the borders of Bavaria and Switzerland. The remainder of the Earth's surface would remain empty of man. Thus the whole of mankind put together represents an insignificant mass of the planet's matter. Its strength is derived not from its matter, but from its brain. If man understands this, and does not use his brain and his work for self-

Portrait by Jean Louis Rodolphe, 1866, courtesy of University of Oklahoma Libraries, History of Science Collections

Louis Agassiz (1807-1873), introduced the idea of the glacial period into science.

NOAA Central Library

Captain Charles Wilkes, headed the U.S. Exploring Expedition, 1838-1842, which discovered the Magnetic South Pole and determined that Antarctica was a continent.

destruction, an immense future is open before him in the geological history of the biosphere.

The geological evolutionary process shows the biological unity and equality of all men, *Homo sapiens* and his ancestors, *Sinanthropus* and others; their progeny in the mixed white, red, yellow, and black races evolves ceaselessly in innumerable generations.[17] This is a *law of nature.* All the races are able to interbreed and produce fertile offspring. In a historical contest, as for instance in a war of such magnitude as the present one, he finally wins who follows that law. One cannot oppose with impunity the principle of the unity of all men as a law of nature. I use here the phrase "law of nature" as this terms is used more and more in the physical and chemical sciences, in the sense of an empirical generalization established with precision.

The historical process is being radically changed under our very eyes. For the first time in the history of mankind the interests of the masses on the one hand, and the free thought of individuals on the other, determine the course of life of mankind and provide standards for mere ideas of justice. Mankind taken as a whole is becoming a mighty geological force. There arises the problem of the *reconstruction of the biosphere in the interests of freely thinking humanity as a single totality.* This new state of the biosphere, which we approach without our noticing, is the *noösphere.*

In my lecture at the Sorbonne in Paris in 1922-1923, I accepted *biogeochemical phenomena* as the basis of the biosphere. The contents of part of these lectures were published in my book, *Studies in Geochemistry,* which appeared first in French, in 1924, and then in a Russian translation, in 1927.[18] The French mathematician Le Roy, a Bergsonian philosopher, accepted the biogeochemical foundation of the biosphere as a starting point, and in his lectures at the Collège de France in Paris, introduced in 1927 the concept of the noösphere as the stage through which the biosphere is now passing geologically.[19] He emphasized that he arrived at such a notion in collaboration with his friend Teilhard de Chardin, a great geologist and paleontologist, now working in China.

The noösphere is a new geological phenomenon on our planet. In it, for the first time, man becomes a *large-scale geological force*. He can, and must, rebuild the province of his life by his work and thought, rebuild it radically in comparison with the past. Wider and wider creative possibilities open before him. It may be that the generation of our grandchildren will approach their blossoming.

How Can Thought Change Material Processes?

Here a new riddle has arisen before us. *Thought is not a form of energy*. How then can it change material processes? That question has not as yet been solved. As far as I know, it was first posed by an American scientist born in Lvov, the mathematician and biophysicist Alfred Lotka.[20] But he was unable to solve it. As Goethe (1740-1832), not only a great poet but a great scientist as well, once rightly remarked, in science we only can know *how* something occurred, but we cannot know *why* it occurred.

As for the coming of the noösphere, we see around us at

Russian Academy of Sciences

The Russian scientist Aleksei Petrovich Pavlov (1854-1929), emphasized that man was becoming a "mighty and ever-growing geological force."

every step the empirical results of that "incomprehensible" process. That mineralogical rarity, native iron, is now being produced by the billions of tons. Native aluminum, which never before existed on our planet, is now produced in any quantity. The same is true with regard to the countless number of artificial chemical combinations (biogenic "cultural" minerals) newly created on our planet. The number of such artificial minerals is constantly increasing. All of the *strategic raw materials* belong here. Chemically, the face of our planet, the biosphere, is being sharply changed by man, consciously, and even more so, unconsciously. The aerial envelope of the land as well as all its natural waters are changed both physically and chemically by man. In the 20th Century, as a result of the growth of human civilization, the seas and the parts of the oceans closest to shore become changed more and more markedly. Man now must take more and more measures to preserve for future generations the wealth of the seas, which so far have belonged to nobody. Besides this, new species and races of

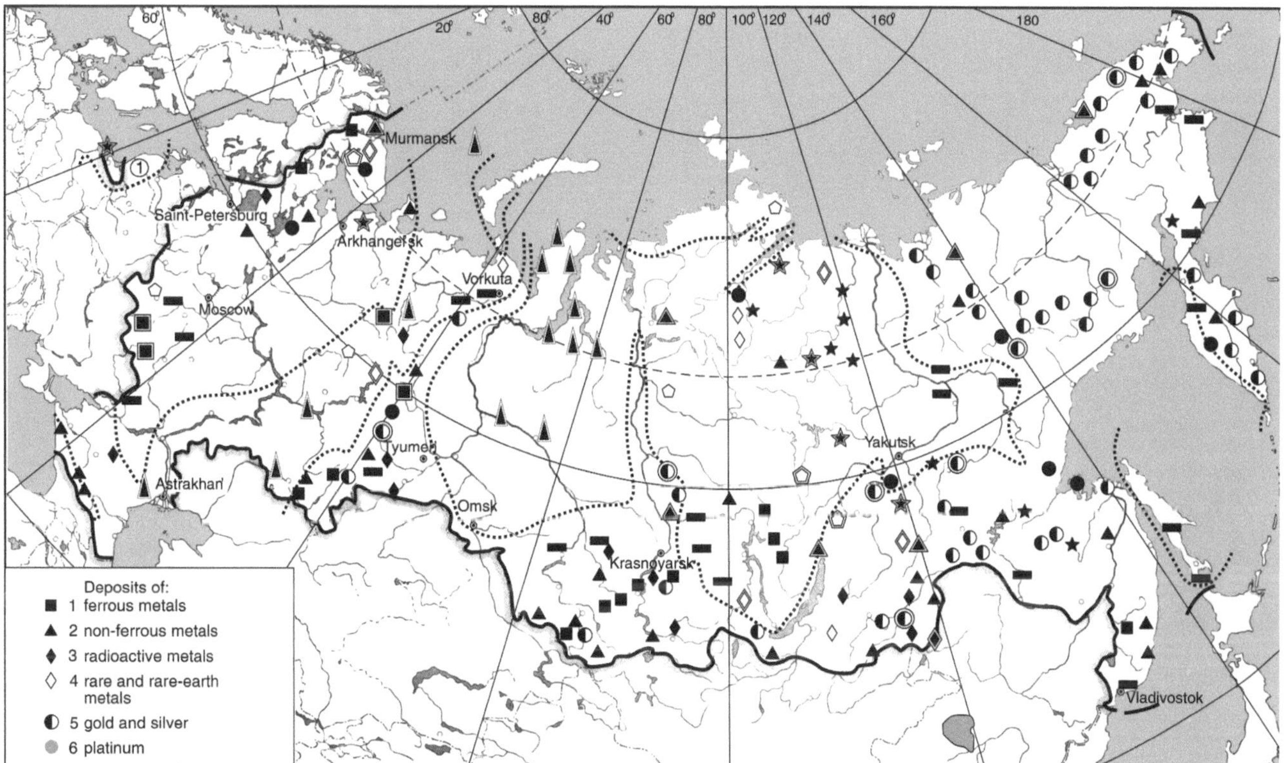

Deposits of:
- ■ 1 ferrous metals
- ▲ 2 non-ferrous metals
- ◆ 3 radioactive metals
- ◇ 4 rare and rare-earth metals
- ◖ 5 gold and silver
- ● 6 platinum
- ⬠ 7 non-metallic ferrous materials
- ★ 8 precious and semi-precious stone
- ◣ 9 oil-and-gas deposits and limits of oil-gas-producing provinces
- ▬ coal

LARGEST MINERAL AND OIL-AND-GAS DEPOSITS OF RUSSIA
The Vernadsky State Geological Museum, Russian Academy of Sciences, has created this map of Russia's resources—to be developed in the interest of mankind: the noösphere.

Source: After Yu. Gatinsky, N. Vishnevskaya, Vernadsky SGMRAS

animals and plants are being created by man. Fairy tale dreams appear possible in the future; man is striving to emerge beyond the boundaries of his planet into cosmic space. And he probably will do so.

At present we cannot afford not to realize that, in the great historical tragedy through which we live, we have elementally chosen the right path leading into the noösphere. I say elementally, as the whole history of mankind is proceeding in this direction. The historians and political leaders only begin to approach a comprehension of the phenomena of nature from this point of view. The approach of Winston Churchill (1932) to the problem, from the angle of a historian and political leader, is very interesting.[21]

The noösphere is the last of many stages in the evolution of the biosphere in geological history. The course of this evolution only begins to become clear to us through a study of some of the aspects of the biosphere's geological past. Let me cite a few examples, Five hundred million years ago, in the Cambrian geological era, skeletal formations of animals, rich in calcium, appeared for the first time in the biosphere; those of plants appeared over 2 billion years ago. That calcium function of living matter, now powerfully developed, was one of the most important evolutionary factors in the geological change of the biosphere.[22] A no less important change in the biosphere occurred from 70 to 110 million years ago, at the time of the Cretaceous system, and especially during the Tertiary. It was in that epoch that our green forests, which we cherish so much, were formed for the first time. This is another great evolutionary stadium, analogous to the noösphere. It was probably in these forests that man appeared around 15 or 20 million years ago.

Now we live in the period of a new geological evolutionary change in the biosphere. We are entering the noösphere. This new elemental geological process is taking place at a stormy time, in the epoch of a destructive world war. But the important fact is that our democratic ideals are in tune with the elemental geological processes, with the law of nature, and with the noösphere. Therefore we may face the future with confidence. It is in our hands. We will not let it go.

Notes _____

1. The word "noösphere" is composed from the Greek terms *noos*, mind, and *sphere*, the last used in the sense of an envelope of the Earth. I treat the problem of the noösphere in more detail in the third part of my book, now being prepared for publication, on *The Chemical Structure of the Biosphere of the Earth As a Planet, and Its Surroundings.*

2. It should be noted that in this connection I came upon the forgotten thoughts of that original Bavarian chemist, C. Schoenbein (1799-1868) and of his friend, the English physicist of genius, M. Faraday (1791-1867). As early as the beginning of the 1840s, Schoenbein attempted to prove that a new division should be created in geology—geochemistry, as he called it. See W. Vernadsky, *Ocherki geokhimii* (Studies in Geochemistry), 4th edition, Moscow-Leningrad, 1934, pp. 14, 290.

3. On the significance of KEPS see A. E. Fersman, *Voina i strategicheskoe syrie* (The War and Strategic Raw Materials), Krasnoufimsk, 1941, p. 48.

4. See my article, "Out of my Recollections: The First Year of the Ukrainian Academy of Sciences," to appear in the Jubilee volume of the *Ukrainian Academy of Sciences* in commemoration of its 25th anniversary.

5. It is to be regretted that the manuscripts left after Wolf's death have been, as yet, neither studied nor published. In 1927, the Commission on the History of Knowledge at the Academy of Sciences decided to do this work, but it could not be accomplished because of the constant changes in the Academy's approach toward the study of the history of science. Now that work at the Academy has been reduced to a minimum, which is harmful to the cause.

6. On the biosphere, see W. Vernadsky, *Ocherki geokhimii*, 4th edition, Moscow-Leningrad, Index; *Biosfera* (The Biosphere), Leningrad, 1926: French edition. Paris, 1929.

7. See my article on "The Geological Envelopes of the Earth as a Planet," *Izvestiia of the Academy of Sciences. Geographical and Geophysical Series*, 1942, p. 251. Cf. H. Spenser Jones, *Life on Other Worlds*, New York, 1940; R. Wildt in *Proc. Amer. Philos. Soc.* 81 (1939), p. 135. A Russian translation of Wildt's study, regrettably not in full (which is not indicated in the paper) appeared in the *Astronomicheskii Zhurnal*, Vol. XVII (1940), No. 5, p. 81ff. By now, a new study by Wildt has appeared, *Geochemistry and the Atmosphere of Planets* (1941), but, to our regret, no copy of it has so far reached us.

8. It would deserve a new edition in modern Russian, with commentaries.

9. See *Ocherki geokhimii*, pp. 9, 288, and my book *Problemy biogeokhimii* (The Problems of Biogeochemistry) III (in press).

10. *Problemy biogeokhimii*, III.

11. In accordance with modern American geologists as, for example, Charles Schuchert (Schuchert and Dunbar, *A Textbook of Geology*, II, New York, 1941, p. 88ff.), I call the Cryptozoic era that period which formerly had been called the Azoic, or the Arcaeozoic, era. In the Cryptozoic era the morphological preservation of the remnants of organisms dwindles almost to nothing, but the existence of life is revealed in the organogenic rocks, the origins of which arouse no doubts.

12. On the bio-inert bodies see W.I. Vernadsky, *Problems of Biogeochemistry*, II, *Trans. Conn. Acad. Arts Sci.*, Vol. 35 (1944), pp. 493-494. Such are, for example, the soil, the ocean, the overwhelming majority of terrestrial waters, the troposphere, and so on.

13. See my basic work referred to in Note 1.

14. See D. Gilman. *The Life of J. D. Dana*, New York, 1899. The chapter on the oceanic expedition in this book was written by Le Conte. Le Conte's book, *Evolution* (1888), has not been accessible to me. His autobiography was published in 1903: W. Armes, Editor, *The Autobiography of Joseph Le Conte*. For his biography and bibliography see H. Fairchild in *Bull. Geol. Soc. Amer.* 26 (1915), p. 53.

15. On Reynolds, see the Index in "Centenary Celebration: Wilkes Exploring Expedition of the U.S. Navy, 1838-1842," *Proc. Amer. Philos. Soc.*, 82, No. 5 (1940). It is to be regretted that our expeditions in the Pacific, so active in the first half of the 19th Century, were later discontinued for a long time (almost until the Revolution), following the death of both Emperor Alexander I (1777-1825) and Count N. P. Rumiantsov (1754-1826)—that remarkable leader of Russian culture who equipped the "Riurik" expedition (1815-1818) out of his private funds.

   In the Soviet period K. M. Deriugin's (1878-1936) expedition should be mentioned; its precious and scientifically important materials have been so far only partly studied and remain unpublished. Such an attitude toward scientific work is inadmissible. The Zoological Museum of the Academy of Sciences must fulfill this scientific and civic duty.

16. D. Gilman, *op.cit.*, p. 255.

17. I and my contemporaries have imperceptibly lived through a drastic change in the comprehension of the circumambient world. In the time of my youth it seemed both to me and to others that man had lived through a historical time only, within the span of a *few thousand years, at best a few tens of thousands of years.* Now we know that man has been consciously living through tens of millions of years. He consciously lived through the glacial period in both Eurasia and North America, through the formation of Eastern Himalaya, and so on. The division of historical and geological time is levelled out for us.

18. The last revised edition of my *Ocherki Geokhimii* (Problems of Geochemistry) appeared in 1934. In 1926, the Russian edition of *Biosfera* (The Biosphere) came out, and in 1929 its French edition. My *Biogeokhimicheskie Ocherki* (Biogeochemical Studies) was published in 1940. The publication of *Problemy biogeokhimii* (Problems of Biogeochemistry) was begun in 1940. (A condensed English translation of Part II appeared, under the editorship of G. E. Hutchinson, in *Trans. Conn. Acad Arts Sci.*, Vol. 35, in 1944.) Part III is in press. *Ocherki geokhimii* was translated into German and Japanese.

19. Le Roy's lectures were at once published in French: *L'exigence idealiste et le fait d'evolution*, Paris, 1927, p. 196.

20. A. Lotka, *Elements of Physical Biology*, Baltimore, 1925, p. 405 ff.

21. W.S. Churchill, *Amid These Storms: Thoughts and Adventures*, New York, 1932, p. 274 ff. I plan to return to this problem elsewhere.

22. I deal with the problem of the biogeochemical functions of organisms in the second part of my book, *The Chemical Structure of the Biosphere*. (see Note 1).

www.ingramcontent.com/pod-product-compliance
Lightning Source LLC
Chambersburg PA
CBHW040755200526
45159CB00026B/2659